Waste to Energy

Waste to Energy

Edited by

S. Syngellakis
Wessex Institute of Technology, UK

WIT*PRESS* Southampton, Boston

Editor:

S. Syngellakis
Wessex Institute of Technology, UK

Published by

WIT Press
Ashurst Lodge, Ashurst, Southampton, SO40 7AA, UK
Tel: 44 (0) 238 029 3223; Fax: 44 (0) 238 029 2853
E-Mail: witpress@witpress.com
http://www.witpress.com

For USA, Canada and Mexico

Computational Mechanics International
25 Bridge Street, Billerica, MA 01821, USA
Tel: 978 667 5841; Fax: 978 667 7582
E-Mail: infousa@witpress.com
http://www.witpress.com

British Library Cataloguing-in-Publication Data

A Catalogue record for this book is available
from the British Library

ISBN: 978-1-78466-060-4
eISBN: 978-1-78466-061-1

Library of Congress Catalog Card Number: 2014957521

Printed by Lightning Source, UK.

Preface

The present volume comprises a collection of articles from the Transactions of the Wessex Institute on the very topical subject of converting the calorific content of waste material into useful forms of energy. It complements and, to a certain degree, overlaps with its companion volume, "*Biomass to Biofuels*",[1] since a significant proportion of biomass converted to energy nowadays originates from various types of wastes.

The material in the first, more substantial part of the volume has been arranged according to the type of process for energy conversion. Biochemical processes are described in six articles, starting with three studies related to the production of methane by anaerobic digestion: reactor conversion efficiencies are assessed in the cases of pork slaughterhouse and landfill leachate, waste sludge and anaerobically degraded sludge as well as kitchen waste and sewage sludge. These assessments are succeeded by investigations on ethanol production from biodegradable municipal solid waste through hydrolysis and fermentation; hydrogen production from glucose through a hybrid anaerobic and photosynthetic process using industrial waste water as substrate; and, finally, biodiesel production from used cooking oil through base-catalyzed transesterification.

Conversions by thermochemical processes are discussed in the subsequent eleven articles of the volume. The most familiar and traditional among them is combustion; the direct use of heat energy for cement kilns is the industrial context within which the combustion properties of meat and bone meal as fuel as well as the effect of various kinds of waste on clinker properties are assessed. Alternatively, the heat produced is used in thermal power stations for steam and, ultimately, electricity generation; municipal solid waste and refuse-derived fuel, provided by refuse companies, are the incinerated feedstock. In another article, computational fluid dynamics

---

[1] S. Syngellakis, ed., *Biomass to Biofuels*, WIT Press, Southampton, 2015, ISBN: 978-1-78466-034-5

modelling is applied to assess the influence of process parameters and to perform optimization studies.

A group of articles deal with more complex thermochemical processes involving combustion combined with pyrolysis and gasification. A study of relevant properties of sewage sludge after anaerobic digestion examines the process efficiency of various feedstocks; reactor performances are experimentally assessed with particular reference to molecular dissociation and the use of waste rubber; sensitivity analyses are performed based on simulations of waste plastics gasification in a fluidized bed reactor.

Two articles focus on biofuels as feed for fuel cells; the first examines the potential of ethanol and glycol recovered from sanitary liquid wastes; the second considers biosyngas from waste biomass and assesses the effect of contaminants on fuel cell performance.

In the last six articles, the emphasis is on management and policy rather than technical issues. Considering manure as biomass for energy conversion, feasibility studies are performed on the technical and economic potential of poultry litter as well as the residual material from pig farms. Finally, several studies on municipal solid waste are concerned with issues affecting its management and concomitant energy technology choices as well as its energy potential based on assessments of its characteristics, accumulation and selective collection.

Stavros Syngellakis
The New Forest, 2015

Acronyms

ASBR	Anaerobic Sequencing Batch Reactor
ABR	Anaerobic Baffled Reactor
AES	Atomic Emission Spectroscopy
ASR	Automotive Shredder Residue
BGP	Brookes Gasification Process
BMSW	Biodegradable Municipal Solid Waste
BOD	Biological Oxygen Demand
CFB	Circulating Fluidised Bed
CHP	Combined Heat & Power
COD	Chemical Oxygen Demand
CRF	Char Reactivity Factor
CSTR	Continuous Stirred-Tank Reactor (model)
CV-AAS	Cold-Vapour – Atomic Absorption Spectroscopy
DAFC	Direct Alcohol Fuel Cell
DM	Dry Matter
DOE	Design of Experiments
DTA	Differential Thermal Analysis
DWVO	Domestic Waste Vegetable Oil
EDX	Energy Dispersive X-ray
EF	Entrained Flow
EIA	Energy Information Administration
ELV	End-of-Life Vehicle
ESP	Electrostatic Precipitator
EWC	European Waste Code
F:M	Food to Microorganism
FAME	Fatty Acid Methyl Ester
FBC	Fluidized Bed Combustor
FFA	Free Fatty Acid
FID	Flame Ionization Detector
FPU	Filter Paper Units
FT	Fischer-Tropsch

FTIR	Fourier Transform Infrared Spectroscopy
GC	Gas Chromatography
GC-MS	Gas Chromatography-Mass Spectrometry
HRSG	Heat Recovery Steam Generator
HRT	Hydraulic Retention Time
ICP	Inductively Coupled Plasma
IGCC	Integrated Gasification Combined Cycle
LHV	Lower Heating Value
MBM	Meat and Bone Meal
MBT	Mechanical Biological Treatment
MFR	Mass Flow Rate
MOGA	Multi-Objective Generic Algorithm
MRD	Minimal Risk Dose
MRF	Material Recovery Facility
MSW	Municipal Solid Waste
NLPQL	Non-Linear Programming by Quadratic Lagrangian
NPDES	National Pollutant Discharge Elimination System
OLR	Organic Loading Rate
OM	Optical Microscopy
PAH	Polycyclic Aromatic Hydrocarbon
PCB	Polychlorinated Biphenyls
PCDD/F	Polychlorinated Dibenzo Dioxins and Furans
PFR	Plug Flow Reactor (model)
POP	Persistent Organic Pollutant
RANS	Reynolds Averaged Navier-Stokes (equations)
RDF	Refuse-Derived Fuel
RMSW	Residual Municipal Solid Waste
RWVO	Restaurant Waste Vegetable Oil
SC	Selective Collection
SGBR	Static Granular Bed Reactor
SMR	Steam Methane Reforming
SRF	Solid Recovered Fuel
TCD	Thermal Conductivity Detector
TG	Thermo-gravimetry
TS	Total Solids
TSS	Total Suspended Solids
UASB	Upflow Anaerobic Sludge Blanket (reactor)
VFA	Volatile Fatty Acids
VS	Volatile Solids
WGS	Water Gas Shift
WTEP	Waste to Energy Plant
WVO	Waste Vegetable Oil
WWTP	Wastewater Treatment Plants
XRD	X-Ray Diffraction
XRF	X-Ray Fluorescence

Contents

High rate, low F:M anaerobic treatment of medium strength waste streams using the static granular bed reactor for renewable energy production

T. G. Ellis[1] & K. M. Evans[2]
[1]*Iowa State University, USA*
[2]*Water and Pollution Control Department, USA*

Abstract

A new high rate, low food to microorganism (F:M) anaerobic treatment system has been developed for maximizing the conversion efficiency of organic matter to energy in the form of methane for medium strength (1–10 g COD L^{-1}) waste streams. The system, termed the static granular bed reactor (SGBR), utilizes a downflow hydraulic regime through a dense bed of active anaerobic granules that can accommodate higher suspended solids concentrations than its counterpart, the upflow anaerobic sludge blanket (UASB) reactor which is susceptible to solids washout due to the high (up to 1 m h^{-1}) upflow velocities. Theoretical considerations for the SGBR are presented with results from SGBR treatment of synthetic (non-fat dry milk and sucrose) and actual (pork slaughterhouse and landfill leachate) wastewaters documenting the high efficiency (consistently greater than 90% conversion of COD) and excellent effluent characteristics (typically less than 50 mg L^{-1} total suspended solids and BOD_5). Transient hydraulic and organic loads have relatively little effect on the SGBR as evidenced by consistent performance during an instantaneous shift in the hydraulic residence time from 36 to 5 h. In addition, results from variable reactor seeding from 25 to 100% of bed volume showed relatively little dependence of granule bed mass on effluent characteristics. This finding confirms independent results showing that a large fraction of the granule bed is substrate limited and available as reserve biomass. These results help explain the excellent treatment performance and resiliency of the SGBR regardless of hydraulic and organic loading. Fatty acid methyl ester (FAME) analysis of the granules showed distinct microbial

communities entirely dependent on the nature of the feed source. Furthermore, FAME profiles from an anaerobic sequencing batch reactor (ASBR) treating pork slaughterhouse wastewater showed distinctly different profiles than an SGBR treating the same wastewater. Energy production from the SGBR was close to the theoretical 0.35 L (g COD removed)$^{-1}$ for all waste streams evaluated.
Keywords: linear alkylbenzene sulfonate, alkyl sulphate, biodegradation, kinetics, competent biomass, respirometry, activated sludge.

1 Introduction

One of the major challenges for high rate anaerobic treatment is the development of a process that can operate over the entire range of ambient temperatures expected in the field and meet the required effluent standard for surface water discharge. Currently, no anaerobic process can achieve this due mainly to the periodic loss of solids from the reactor especially as the system experiences hydraulic fluctuations. The upflow anaerobic sludge blanket (UASB) reactor, for instance, uses an upflow configuration with a three phase (solids/liquid/gas) separator at the top. Due to the upflow nature of the process and the fact that anaerobic sludge granules and flocculent sludge tends to float, UASB reactors are particularly susceptible to washout of solids [1–4].

The three phase separator in the UASB must collect the gas that is produced in the reactor, clarify the effluent, and keep the granular sludge in the reactor. The way this is accomplished is through the use of baffles and a gas collector. The gas bubbles are directed to the gas collector by perimeter baffles and effluent has to pass around the baffles and out through the effluent weir. The intent is that the granular biomass will settle back to the sludge bed. There are several difficulties with this arrangement. The gas that is produced by the anaerobic granules is in the form of small bubbles at the surface of the granule. As the bubble size increases due to additional microbial activity, surface tension keeps the bubble on the granule exterior until the buoyancy force exceeds the surface tension. At the same time, the buoyancy of the granule changes due to the increased surface area and the change in density due to the attached bubbles. The only downward force on the granule is the force of gravity. Upflow forces include buoyancy forces and the upflow velocity. Depending on the movement of the granule, there may also be drag forces counteracting the gravitational or buoyancy forces. At steady state operating conditions, maintaining the anaerobic granules in the UASB reactor requires that the force of gravity exceed the buoyancy and upflow forces. During hydraulic (influent flowrate) and temperature changes within the UASB, however, this becomes increasingly difficult.

Two things happen to the granule during an increase in the hydraulic loading in a UASB. The first is that the organic loading (mass of organic matter per volume per time) increases due to the increased flowrate. Consequently, the granules tend to produce more gas since more substrate is available. The additional gas production increases the buoyant force on the granule. Secondly, the upflow velocity increases due to the increased flowrate. As the forces from buoyancy and the upflow velocity become greater than the gravitational force, the granule rises to the top of the reactor. Depending on the efficiency of the gas/solids/liquid separator, the buoyant granule may be washed out of the reactor with the effluent. This same effect happens to flocculent biomass, but its loss is not necessarily detrimental to the UASB, since the granular biomass is preferred. Loss of either granular or flocculent biomass, however, significantly increases the effluent solids concentration. Temperature shifts tend to exasperate this effect since the changes in liquid density and biological activity are temperature dependent and result in varied combinations of buoyancy, upflow and gravitational forces.

To circumvent these problems, a new reactor configuration, called the static granular bed reactor (SGBR) was developed by researchers at Iowa State University [5–13]. The SGBR utilizes a bed of active anaerobic granules in a downward flow regime (see fig. 1). The innovation in this reactor configuration is that it uses the highly active anaerobic granules (just as in a UASB system), but it operates in a downflow mode. The advantage of a downflow configuration is that the biogas that is generated rises and is easily separated from the granules and the liquid at the top of the reactor. Granule buoyancy is not a detriment to process performance in the SGBR as it is in the UASB. In contrast to the UASB, there is no need for a sophisticated three phase solids, gas and liquid separator. Neither is there a need for recirculation pumps, timers, mixers, or other ancillary equipment that are required for the UASB systems. Consequently, the effluent quality of the SGBR is improved in comparison to the UASB. The biomass granules are retained within the reactor by the use of a gravel underdrain and temperature and hydraulic loading changes are not expected to significantly affect effluent quality.

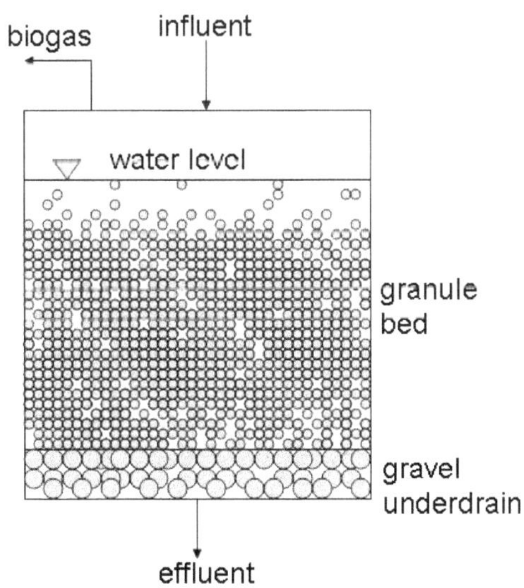

Figure 1: Schematic of the Static Granular Bed Reactor (SGBR). US patent #6,709,591.

The technological innovation of the SGBR is that it uses highly active anaerobic granular biomass in a downflow configuration. Other reactor configurations use a downward flow regime (e.g., the anaerobic filter can be operated in a downflow mode), but the SGBR is the first system to use anaerobic granules in a downflow mode. This configuration allows for an exceptional effluent quality, simple operating characteristics, and smaller volume requirements. In addition, there is no need to heat the SGBR in most cases, since the high retention of biomass ensures that there is sufficient microbial activity to degrade the organic portion of the influent wastewater, allowing more of the energy produced in the form of methane in the biogas to be available for productive uses.

Laboratory and pilot-scale SGBR studies have demonstrated the ability of the system to treat a wide variety of waste streams including pulp and paper wastewater [14], industrial wastewater from dairy processing [15], swine manure [16–17], pork slaughterhouse wastewater [18], and synthetic wastewater [19].

2 Materials and methods

Laboratory scale SGBRs were constructed from Plexiglas cylinders with working volumes from 1 to 10. A stainless steel mesh (15 mm) or pea gravel was placed in the bottom of the SGBRs to retain the granular biomass. A "T" connector was installed in the effluent line to maintain the desired working volume in the reactor. Seed granules were obtained from an operating UASB at CityBrew (formerly Heileman's Brewery) in LaCrosse, WI. Synthetic feed consisting of nonfat dry milk and nutrients was fed on a semi-continuous basis using a peristaltic pump. Feed and effluent were both stored at 4°C. Analytical parameters (alkalinity, COD, suspended solids) were determined using Standard Methods and performed weekly. The VFA concentration was measured by the titration method (Method 5560 C in Standard Methods) with an assumed efficiency of 70%. Gas composition was analyzed by a Gow Mac gas chromatograph and was tested bi-weekly. BOD_5 concentration was measured once during each HRT condition. In the first series of experiments, two different SGBR reactor height:width configurations were evaluated. The first reactor (SGBR 1) had an inner diameter of 101 mm and a granule height of 135 mm. SGBR 2 had an inner diameter of 64 mm and a granule height of 432 mm. Both were operated for approximately two years under several HRT conditions at ambient temperature (22 ± 2°C). In the next set of experiments, four 1-L SGBRs were seeded with varying amounts of seed granules (250, 500, 750, and 1,000 mL) to evaluate the optimum granule bed depth as a percentage of working volume. An evaluation of whole cell fatty acid methyl esther (FAME) was used to assess the microbial community structure similarities between SGBRs treating different wastewaters and other reactors treating the same wastewater. FAMEs were extracted using the Microbial Identification System anaerobic extraction protocol (Microbial ID, Inc., Newark, DE). Cells in the samples were saponified by heat and the presence of a strong base. In this step, fatty acids were separated from lipids. After the separation, the remaining FAs were methylated to form FAME and extracted into an organic solvent. Following extraction, FAME samples were analyzed on a HP 6890 (Hewlett Packard, Rolling Meadows, IL) gas chromatograph. Because of the large number of FAs present in each sample, principal component (PC) analysis was used to observe any changes in the FAME profiles. Principal component analysis explains the maximum variation in the data based on theoretical components, or principal components in the data. For this study, PC analysis was done by MIDI FAME's (Microbial ID, Inc., Newark, DE) Sherlock program.

3 Results and discussion

During all HRT conditions both SGBRs performed well with respect to COD and BOD_5 removals and the effluent from both had low effluent VFA, BOD_5, and TSS concentrations (see table 1). The poorest performance occurred at the initial start-up condition (36 h HRT) for SGBR 1. This could be due to the limited acclimation and washout of fine solids during start-up. After initial start-up, SGBR 1 and 2 maintained greater than 92% and 94% total COD removal respectively. Soluble COD removal in both was 95–97%.

Table 1: SGBR performance characteristics as a function of reactor dimensions.

SGBR 1 HRT	Effluent COD, mg/L	Effluent BOD, mg/L	TSS, mg/L	VFA, mg/L as CH_3COOH	H_2S in gas, Ppm
Start-up 36	523±120	NA	130	24±5	NA
24	45±28	10	20	14±5	650
16	30±15	10	6	14±6	600
12	30±19	26	16	18±7	600
8	79±29	47	8	17±2	1800
6	38±24	27	10	19±1	1400
5	39±15	22	12	17±5	120
36	46±18	13	8	16±2	600
5	40±22	10	19	9±1	300
SGBR 2					
Start-up 36	52±23	8	5	12±3	1700
24	59±15	NA	18	17±5	500
16	44±24	26	3	17±5	NA
12	50±21	21	6	17±1	850
8	33±15	14	5	14±1	600
5	24±12	12	7	12±5	800
36	41±19	10	5	11±1	1100
5	41±18	9	10	9±1	500

Throughout the research period, SGBR 2 consistently had lower measured VFA concentrations. Gas chromatographic (GC) analysis, however, indicated that the titration method might have overestimated the VFA concentration. By GC analysis, acetic, propionic, butyric, and valeric acid concentrations were each measured at or below the detection limit of 1 mg/L. Suspended solids and BOD_5 consistently met (with the exception of the 8 hour HRT condition for SGBR 1) the NPDES requirement of 30 mg/L which would allow the effluent to be discharged to surface water in areas where nutrient removal was not required.

In order to evaluate the organic loading turndown ratio in the SGBR, a hydraulic and organic loading transition was initiated by changing the HRT from 36 to 5 h. The transition was instantaneous and analytical testing was performed every 12 h to observe the non-steady state performance (see fig. 2). After the HRT was lowered (resulting in a corresponding increase in organic loading), the effluent total COD increased from 46 mg/L to 144 mg/L in SGBR 1 (the wider and shorter reactor). However, SGBR 2 showed no significant increase in effluent COD concentration. Both reactors showed an increase in TSS, but the concentration for both remained below 30 mg/L. Soluble COD increased initially, but returned to low levels (< 30 mg/L) within 36 h. Other parameters such as the VFA concentration were not significantly affected. The VFA concentration remained low (< 12 mg/L by titration method) during the transition period. The hydrogen sulfide concentration remained less than 150 ppm during the transition period, which was lower than that measured during the previous HRT conditions (table 2).

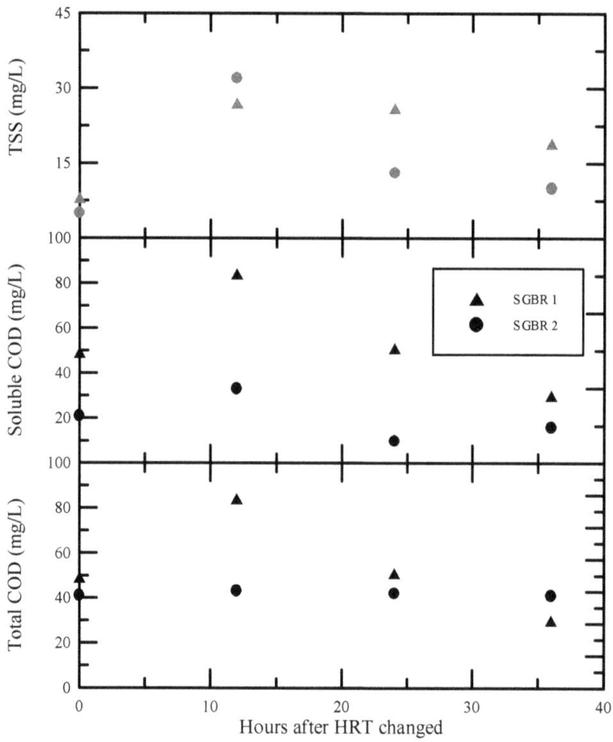

Figure 2: Effluent TSS and COD during transition period from 36 to 5 h HRT.

The operation at the 5 h HRT remained exceptional. The organic removal remained high. COD removal in SGBR 2 improved slightly, but not significantly (95% at 36 h to 97% at 5 h). Gas production increased with the higher organic loading and the methane content was 81% for both SGBRs.

Table 2: Effluent characteristics as a function of granule seeding.

Bed fraction	1 g COD L^{-1}		2 g COD L^{-1}		4 g COD L^{-1}	
	COD	TSS	COD	TSS	COD	TSS
25%	28 ± 17	26 ± 14	42 ± 22	40 ± 15	304 ± 153	146 ± 12
50%	33 ± 22	24 ± 13	35 ± 21	26 ± 14	1121 ± 456	184 ± 8
75%	22 ± 14	21 ± 7	15 ± 11	25 ± 10	1043 ± 534	113 ± 9
100%	14 ± 17	18 ± 8	34 ± 27	20 ± 7	294 ± 176	80 ± 11

In a parallel study, the microbial community in the SGBR was analyzed using FAME analysis. Two parallel reactors (an SGBR and an anaerobic sequencing batch reactor) treating slaughterhouse wastewater (1.9 g COD L^{-1}) were compared along with an SGBR treating non fat dry milk (1 g COD L^{-1}). The FAME profiles and corresponding comparisons using principal component analysis indicate the similarities or differences in the fatty acid (FA) content of each sample. These profiles can be correlated to the microbial population during the different sampling times and conditions. From the principal component plot, distinct groupings of fatty acids profiles suggest homogeneity of microbial communities (see fig. 3). The ASBR had a slightly more homogenous population profile which is consistent with the mixing and settling during the operation of the system, while the SGBR profile was more diverse.

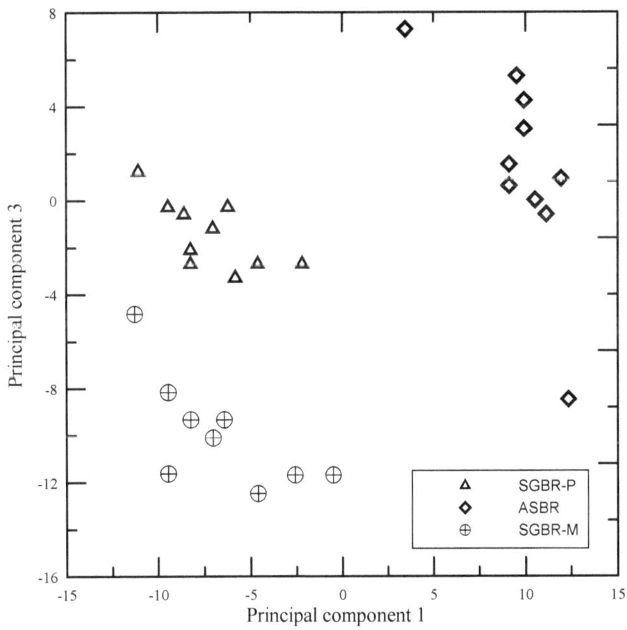

Figure 3: Two dimensional principal component analysis of FAME profiles (SGBR-P and ASBR were fed pork slaughterhouse and SGBR-M was synthetic dairy wastewater).

To further evaluate the performance characteristics of the SGBR, a number of studies have been performed on synthetic and actual wastewaters and the detailed results have been presented elsewhere. A summary of these studies is useful, however, for comparison purposes. Fig. 4 shows the performance of the SGBR [10, 20] as a function of OLR in comparison to UASB [22, 23], anaerobic filter [23], and expanded sludge bed [24] reactors treating either slaughterhouse or landfill leachate wastewater. In all cases the response of the SGBR to an increase in OLR is relatively flat (i.e., unaffected by OLR increase), whereas the other high rate reactor configurations suffered noticeable decline in performance.

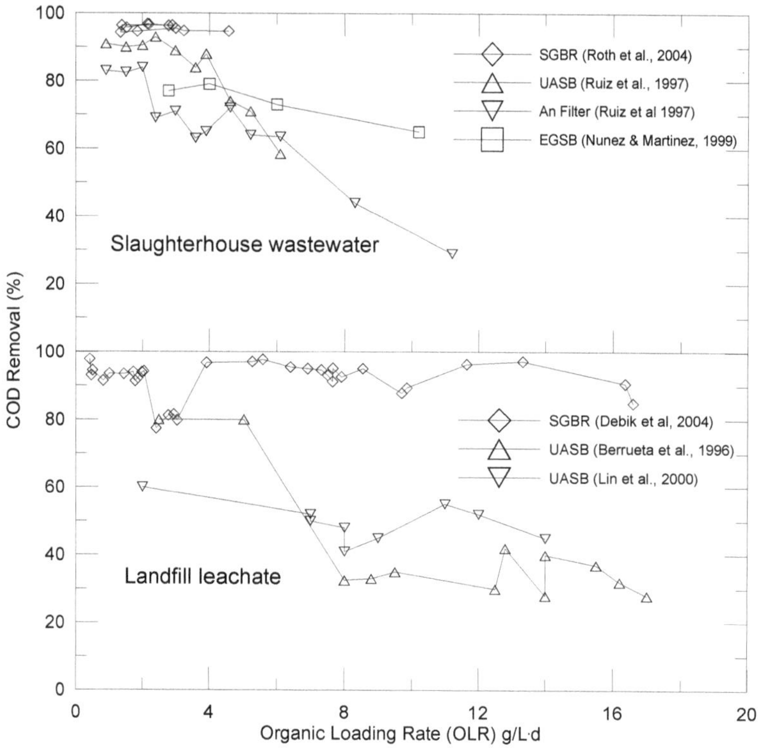

Figure 4: Performance characteristics of high rate anaerobic treatment systems as a function of OLR and wastewater type.

The response of the SGBR can be related to the manner in which the SGBR retains solids in the reactor resulting in extremely low food:microorganism (F:M) ratios regardless of OLR. Because of the high retention of solids in the SGBR (and resulting low F:M), a potential failure in the SGBR would be more likely due to hydraulic limitations than organic overloading [14]. Consequently, careful attention to bed cleaning through backwashing is critical to maintain the hydraulic capacity of the reactor. A recent study on dairy processing wastewater showed that total backwash volumes of less than 5% of treated water were sufficient to maintain the hydraulic capacity of the 42.5 m^3 (working volume) pilot reactor [15].

4 Conclusion

From these studies, it can be seen that the SGBR holds promise as an innovative bioreactor design to broaden the range of substrates for high rate, low F:M anaerobic digestion and concomitant conversion of organics to renewable fuel in the form of methane. The SGBR's unique configuration retains the active granular biomass in the reactor regardless of OLR and hydraulic loading and loading rate changes, resulting in extremely low F:M ratios. This allows the SGBR an advantage over other high rate anaerobic treatment technologies which may be susceptible to solids washout, loss of granular biomass and subsequent process deterioration.

References

[1] Goodwin, J.A.S., Finlayson, J.M. & Low, E.W., A further study of the anaerobic biotreatment of malt whisky distillery pot ale using an UASB system. *Bioresource Technology*, **78**, pp. 155-160, 2001.

[2] Dinsdale, R.M., Hawkes, F.R. & Hawkes, D.L., Comparison of mesophilic and thermophilic upflow anaerobic sludge blanket reactors treating instant coffee production wastewater. *Water Research*, **31**, pp. 163-169, 1997.

[3] Oktem, Y.A., Ince, O., Sallis, P., Donnelly, T. & Ince, B.K., Anaerobic treatment of a chemical synthesis-based pharmaceutical wastewater in a hybrid upflow anaerobic sludge blanket reactor. *Bioresource Technology*, **99**, pp. 1089-1096, 2007.

[4] Punal, A., Lorenzo, A., Roca, E., Hernandez, C. & Lema, J.M., Advanced monitoring of an anaerobic pilot plant treating high strength wastewaters. *Water Science and Technology*, **40(8)**, pp. 237-244, 1999.

[5] Mach, K.F. & Ellis, T.G., Height to width ratio comparisons of the static granular bed reactor. *Proc. of the 73rd Annual Water Environment Federation Conference & Exposition*, Anaheim, California, USA, 2000.

[6] Mach, K.F. & Ellis, T.G., Fatty acid methyl ester (FAME) analysis of high rate anaerobic wastewater treatment systems. *Proc. of the 74th Annual Water Environment Federation Conference & Exposition*, Chicago, Illinois, USA, 2002.

[7] Mach, K.F. & Ellis, T.G., Performance evaluation of two static granular bed reactors. *Proc. of the 9th World Congress on Anaerobic Digestion*, Antwerpen, Belgium, 2001.

[8] Mach, K.F. & Ellis, T.G., Fundamentals of the static granular bed reactor. *Proc. Asian Waterqual 2003*, IWA-Asia Pacific Regional Conference, Bangkok, Thailand, 2003.

[9] Evans, E.A. & Ellis, T.G., Analyzing an anaerobic alternative: A pilot-scale study of a static granular bed reactor. *Water Environment and Technology*, **18(3)**, pp. 69-73, 2006.

[10] Debik, E., Park, J. & Ellis, T.G., Leachate treatment using the static granular bed reactor. *Proc. of the 77th Annual Water Environment Federation Conference & Exposition*, Washington, D.C., USA, 2005.

[11] Evans, E.A. & Ellis, T.G., Anaerobic municipal wastewater treatment using the static granular bed reactor. *Proc. of the 77th Annual Water Environment Federation Conference & Exposition*, Washington, D.C., USA, 2005.
[12] Evans, E.A. & Ellis, T.G., Industrial wastewater treatment with the static granular bed reactor versus the UASB. *Industrial Wastewater*, **5(6)**, pp. 10, 2007.
[13] Ellis, T.G., Park, J., Debik, E., & Smith, S., Evaluation of leachate treatment and recycle options at the Boone County Landfill. *Proc. of the 20th Annual Solid Waste Management Conference*, Philadelphia, PA, USA, 2005.
[14] Ilter, F.; Park, J.; Evans, E. A.; Ellis, T.G., Evaluation of pretreatment using UASB and SGBR reactors for pulp and paper plants wastewater treatment. *Water, Air & Soil Pollution*, **224**, pp. 1512, 2013.
[15] Park, J., Oh, J.H., Evans, E.A., Hobson, K., Lally, M. & Ellis, T.G., Industrial wastewater treatment by on-site pilot static granular bed reactor (SGBR). *Water Practice and Technology*, **7(1)**, pp. 2166, 2012.
[16] Lim, S.J., Fox, P. & Ellis, T.G., Corrigendum to "A kinetic evaluation of anaerobic treatment of swine wastewater at two temperatures in a temperate climate zone". *Bioresource Technology*, **144**, pp. 698, 2013.
[17] Lim, S.J., Fox, P. & Ellis, T.G., Corrigendum to "Evaluation of a static granular bed reactor using a chemical oxygen demand balance and mathematical modelling". *Bioresource Technology*, **144**, pp. 699, 2013.
[18] Park, J., Oh, J.H. & Ellis, T.G., Evaluation of an on-site pilot-scale static granular bed reactor (SGBR) for the treatment of slaughterhouse waster. *Bioprocess Biosystems Engineering*, **35**, pp. 459-468, 2012.
[19] Evans, E.A. & Ellis, T.G., Experimental validation of the static granular bed reactor for industrial waste anaerobic treatment. *Journal of Environmental Engineering*, **136**, pp. 1139-1146, 2010.
[20] Roth, M.J. & Ellis, T.G., Onsite pilot demonstration of the static granular bed reactor. *Proc. of the 76th Annual Water Environment Federation Conference & Exposition*, New Orleans, Louisiana, USA, 2004.
[21] Berrueta, J., Gutierrez, A. & Fueyo, G., Anaerobic treatment of leachates in a pilot scale UASB: Strategy of start-up. *Journal of Chemical Technology*, **67**, pp. 302-314, 1999.
[22] Lin, C.Y., Chang, F.Y. & Chang, C.H., Co-digestion of leachate with septage using a UASB reactor. *Bioresource Technology*, **68**, pp. 275-282, 2001.
[23] Ruiz, I., Veiga, M.C., de Santiago, P. & Blazquez, R., Treatment of slaughterhouse wastewater in a UASB reactor and anaerobic filter. *Bioresource Technology*, **60**, pp. 251-258, 1997.
[24] Nunez, L.A. & Matinez, B., Anaerobic treatment of slaughterhouse wastewater in an expanded granular sludge bed (EGSB) reactor. *Water Science and Technology*, **40(8)**, pp. 99-106, 1999.

Squeezing wastes in a wastewater treatment plant

I. M. Buendía, F. J. Fernández, J. Villaseñor & L. Rodríguez
University of Castilla-La Mancha, Spain

Abstract

In this study, the performance of an anaerobic digester placed in a conventional wastewater treatment plant was evaluated. For that purpose, two different kinds of sludge were studied: a mixture of primary and secondary sludge, i.e. waste sludge, and the sludge after being anaerobically degraded, i.e., stabilized sludge. The practical biogas potential of the two wastes was studied in batch anaerobic reactors under mesophilic conditions and was compared with the theoretical methane potential that was expected from the analyses of different organic fractions of the two wastes. The obtained results indicated that there was approximately a 30% of biogas potential without recovering when the waste sludge was treated in the anaerobic digester. The theoretical and practical methane potential of the stabilized sludge, 0.16 L CH_4 g VS^{-1} and 0.33 L CH_4 g VS^{-1} respectively, showed that there were still organic fractions in the stabilized sludge that could be anaerobically degraded. That means that it is possible to enhance the methane production of the anaerobic digester of a conventional wastewater treatment plant by changing some operational conditions, for instance, the hydraulic retention time.
Keywords: anaerobic digestion, waste sludge, wastewater treatment plant.

1 Introduction

The rapid increase of population in recent years has resulted in the generation of a large amount of organic wastes. That makes it necessary to change the methodology for treating those wastes or to improve the existing infrastructure.

One of the most important organic wastes generated in the industrialised countries is the sludge produced in the wastewater treatment plants (WWTPs). In Spain, the production of the sludge generated in the WWTPs has increased by 39 percent within the period from 1997 to 2005. For example, in the region of Castilla-La Mancha (Spain), the production of sludge was 42,917.51 tons [1] in 2003. In the old days, the sludge was directly dumped to the landfills. Currently, there are specific regulations for avoiding landfill disposal of the sludge generated in the WWTPs or for minimising its dump, at the same time, these regulations are mainly focused on the energetic valorisation of that sludge [1, 2].

The stabilization of the sludge is mainly carried out by its own WWTP. The most commonly used technology in the WWTP to stabilize this organic waste is the anaerobic digestion, since organic matter has been acknowledged as a source of renewable energy as the anaerobic decomposition process generates fuel biogas containing 55 to 75% of methane. In addition, using anaerobic biological treatments to stabilized the sludge, an organic residue is produced that can be used as a soil-beneficial substrate, i.e., as a fertilizer. It has been reported that a hygienic stabilization step is sometimes needed for subsequent land application [3, 4].

Some studies suggest the improvement of the anaerobic digestion of the sludge generated in a WWTP by co-digesting the sludge with other organic wastes [5, 6] or by pre-treating the sludge with mechanical, thermal or chemical pre-treatments prior to its digestion [7–9]. However, the low performance of the anaerobic digesters located at the conventional WWTPs is commonly attributed to some technical problems or to their oversize [10].

In this context, the aim of this work was to study both the practical and theoretical methane potential of the waste sludge and the stabilized sludge generated in a WWTP in order to evaluate the success of the anaerobic digester of the WWTP.

2 Material and methods

2.1 Inoculum

Mesophilically digested sludge from the anaerobic digester of a conventional activated sludge wastewater treatment plant was used as inoculum for the anaerobic experiments. The anaerobic inoculum was kept at 35°C during two weeks in order to ensure that all biodegradable organic fractions were consumed.

2.2 Substrate source

Waste sludge and stabilized sludge from a conventional activated sludge wastewater treatment plant were used as organic substrates.

The waste sludge was a mixture of both primary and secondary sludge. The stabilized sludge was previously digested in an anaerobic digester, which operated at 20 days of hydraulic retention time (HRT) in the same wastewater treatment plant.

2.3 Anaerobic experiments

2.3.1 Calculation of theoretical methane potential

The stabilization of the organic wastes by anaerobic degradation is directly related to both methane and carbon dioxide production, thus the formula for predicting the methane and carbon dioxide production from a knowledge of the waste chemical composition is as following [11, 12]:

$$C_nH_aO_b + \left(n - \frac{a}{4} - \frac{b}{2}\right)H_2O \rightarrow \left(\frac{n}{2} + \frac{a}{8} - \frac{b}{4}\right)CH_4 + \left(\frac{n}{2} - \frac{a}{8} + \frac{b}{4}\right)CO_2 \quad (1)$$

The knowledge of the atomic composition of the most organic wastes is sometimes very difficult to achieve and it is necessary to take into account the analytical organic fractions. Therefore, the theoretical methane potential ($\beta_{o,th}$) expressed as litres of CH_4 in standard conditions of temperature and pressure per gram of volatile solids (VS) added, can be assessed as follows [13]:

$$\begin{aligned}\beta_{o,th}\ (\text{STP L}\,CH_4\ \text{g}\,VS^{-1}) &= 0.415\times\text{carbohydrates} + 0.496\times\text{proteins} + \\ &+ 1.014\times\text{lipids} + 0.373\times\text{acetate} + 0.530\times\text{propionate}\end{aligned} \quad (2)$$

where the organic fractions (carbohydrates, proteins, lipids, acetate and propionate) are expressed as % of the total VS, and the coefficients are obtained from stoichiometric conversion of the different organic fractions according to eqn (1).

2.3.2 Biogas potential experiments

For the study of the biogas potential of the different organic wastes, anaerobic experiments were carried out in 1 L-batch reactors with a working volume of 400 mL. In the wall of the reactors two sampling valves were fitted to take liquid and gas samples of the liquid phase and gas phase, respectively. Additionally, a pressure sensor was incorporated at the top of the reactor to measure the biogas pressure in the gas phase. The reactors were placed in an incubator, which had a temperature control for maintaining the temperature at 35°C. A magnetic stirred ensured that solids remain in suspension and a good degree of contact between the organic wastes and the microorganism population contained in the inoculum. At the beginning of the experiments, a mixture of nitrogen/carbon dioxide gas was flushed during 15 minutes into the reactors for ensuring anaerobic conditions.

The food/microorganism (F/M) ratio was set in at 1 g VS g VS^{-1}. In addition, reference experiments were carried out to take into account the endogenous biogas production.

Cumulative methane production (β) from the wastes was assessed from experimental biogas production and methane content data after subtracting the amount of biogas generated by the reference experiments.

2.4 Analytical methods

Methane and carbon dioxide concentrations in gas samples were determined with gas chromatography using a thermal conductivity detector (TCD). For the quantification of volatile fatty acids (VFA): acetate, propionate, i-butirate, n-butirate, i-valerate and n-valerate; acidified samples with 34% H_3PO_4 were analysed on a gas chromatography using a flame ionization detector (FID). Total Solids (TS), volatile solids (VS), ammonia-nitrogen (NH_4^+-N), total-phosphorus (T-P), lipids and pH were determined according to standard methods [14]. Heavy metals were measured using an ICP-Mass Spectrometry.

3 Results and discussion

3.1 Characterization of the organic wastes

Prior to the anaerobic experiments, an exhaustive characterization of the organic wastes was needed with the aim of determining the different organic fraction (carbohydrates, proteins, lipids, acetate and propionate) contained in each waste.

The characterization of the waste sludge and stabilized sludge is shown in table 1.

Table 1: Characterization of the wastes.

Parameters	Waste sludge	Stabilized sludge
TS (%)	4.12 ± 1.05	2.02 ± 1.72
VS (%)	3.84 ± 0.82	1.23 ± 0.70
pH	6.40 ± 0.20	7.00 ± 0.30
Total-P (mg L^{-1})	920 ± 83	1,020 ± 302
NH_4^+-N (mg L^{-1})	290 ± 12	340 ± 103
Lipids (mg L^{-1})	4,100 ± 350	3,300 ± 421
Acetate (mg L^{-1})	2,941 ± 120	1,600 ± 582
Propionate (mg L^{-1})	1,300 ± 47	860 ± 274
Proteins (mg L^{-1})	1,812 ± 121	2,125 ± 230
Carbohydrates (mg L^{-1})	31,247 ± 500	4,415 ± 117
Cd (mg kg^{-1}) *	0.51 ± 15	< det.lim
Cr (mg kg^{-1})*	46 ± 7	21 ± 7
Cu (mg kg^{-1})*	231 ± 42	310 ± 13
Pb (mg kg^{-1})*	48 ± 21	< det.lim
Ni (mg kg^{-1})*	89 ± 19	25 ± 13
Hg (mg kg^{-1})*	0.03 ± 0.01	< det.lim

*Composition given in dry matter; det.lim: detection limit.

In table 1, the content of acetic acid and propionic acid is only shown, since the amount of the rest of VFA analysed (i-butirate, n-butirate, i-valerate and n-valerate) was practically negligible. Although the values of acetate and propionate in stabilized sludge were high compared to the values found in the literature [16], the performance of the anaerobic digestion process was high with regards to the organic solids removal achieved, since TS removal and VS removal after the anaerobic treatment of the waste sludge were approximately 51% and 68%, respectively.

The results of the characterization indicate that both waste sludge and stabilized sludge could be used for agricultural purposes according to the content of the heavy metals analysed [15].

3.2 Biogas potential experiments

Anaerobic experiments under mesophilic conditions were performed in triplicate in order to determine the practical biogas potential (β) of the two sludges subject to study: waste sludge and stabilized sludge (fig. 1).

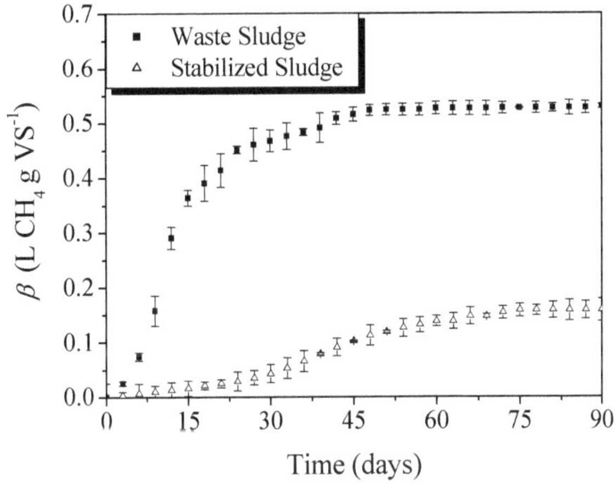

Figure 1: Cumulative methane production per gram of VS loaded experiments of the different wastes.

The length of the anaerobic experiment was 90 days to ensure that all biodegradable organic fractions of the wastes were totally degraded. Regarding fig. 1, a lag phase could not be observed during the digestion of the waste sludge. That could be explained as the inoculum used for carrying out the anaerobic experiments was taken from the anaerobic digester of the WWTP where the waste sludge was generated, therefore the inoculum was acclimated to this kind of waste.

During the first 20 days of digestion of the waste sludge, the methane potential achieved was 0.37 L CH_4 g VS^{-1}. At the end of the experiments, the methane potential reached a value of 0.53 L CH_4 g VS^{-1}. Taking into account that the anaerobic digester of the WWTP operated at 20 days of HRT, these results show that there was about a 30% of methane potential without recovering when the waste sludge was treated under anaerobic conditions in that WWTP, a value that was coincident with the methane potential achieved when digesting the stabilized sludge, 0.16 L CH_4 g VS^{-1}. That means that there was still an organic fraction in the stabilized sludge that could be anaerobically degraded.

Fig. 2 shows the evolution of the VFA concentration during the anaerobic digestion of the different wastes.

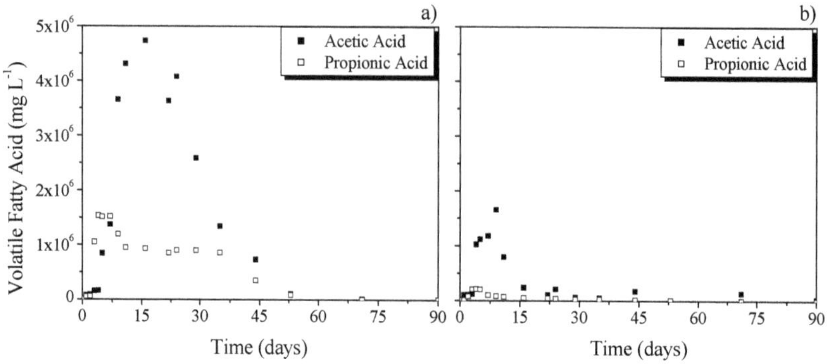

Figure 2: Acetic and propionic acid production. a) Waste sludge; b) Stabilized sludge.

High levels of acetate and propionate were noticed in comparison with the concentration of the rest of VFA (i-butirate, n-butirate, i-valerate and n-valerate); these values were negligible, therefore they are not shown in fig. 2.

A typical behaviour in VFA production was observed during the batch anaerobic degradation of the biodegradable organic fraction of the two wastes, since VFA are intermediate reaction products when the degradation of the complex organic polymers (lipids, carbohydrates, proteins) takes place.

A higher production of acetate and propionate was observed during the digestion of waste sludge due to the high organic matter present in the waste sludge in contrast with the organic content of the stabilized sludge (table 1).

3.3 Comparison of theoretical and practical methane potential

Table 2 shows the theoretical methane potential of both waste sludge and stabilized sludge at different temperature and pressure conditions. The practical methane potential achieved during the anaerobic degradation of each waste is also shown in table 2.

Table 2: Theoretical methane potential ($\beta_{0,th}$) and practical methane potential (β) of the different wastes.

Waste type	$\beta_{0,th}$ (STP) LCH_4 gVS^{-1}	$\beta_{0,th}$ (35°C, 1 atm) LCH_4 gVS^{-1}	β (35°C, 1 atm) LCH_4 gVS^{-1}
Waste Sludge	0.48	0.55	0.53
Stabilized Sludge	0.29	0.33	0.16

In table 2, it can be observed that the practical methane potential was lower than the theoretical methane potential for both wastes under the same temperature and pressure conditions.

For the assessment of the theoretical methane potential was necessary to know the waste composition (table 1), and based on that composition, an approximate calculation of the expected biogas production could be achieved. Nevertheless, the practical methane potential obtained in an anaerobic reactor is always lower because, for instance, some part of the organic matter is used to synthesize new biomass or a part of the organic matter cannot be degraded due to its structure, which makes impossible the access of the anaerobic bacteria to the organic matter [13].

As expected, the methane potential (theoretical and practical potential) of the stabilized sludge was much lower than that obtained for the waste sludge, since the stabilized sludge was previously digested.

4 Conclusions

The main conclusions of this study are shown below:

- The anaerobic digestion process had a high efficiency, with a VS removal of 68%.
- Both sludges could be used for agricultural purposes according to the content of the heavy metals analysed.
- There was about a 30% of methane potential without recovering when the waste sludge was treated in the anaerobic digester of the WWTP.
- It is possible to enhance the methane production of an anaerobic digester of a conventional WWTP by changing some operational conditions such as the hydraulic retention time.

References

[1] Decree 32/2007. Decreto 32/2007, de 17 de abril de 2007, por el se aprueba el plan de gestión de los lodos producidos en las estaciones depuradoras de aguas residuales de Castilla la Mancha. *Spanish Decree*, 2007.

[2] PNIR 2007-2015. II Plan nacional de lodos de depuradoras de aguas residuales – EDAR II PNLD (2007–2015). *Spanish Regulation*, 2007.

[3] Sosnowski, P., Wieczorek, A. & Ledakowicz, S., Anaerobic co-digestion of sewage sludge and organic fraction of municipal solid wastes. *Advances in Environmental Research,* 7, pp. 609–616, 2003.

[4] Watanabe, H., Kitamure, T., Ochi, S. & Ozaki, M., Inactivation of pathogenic bacteria under mesophilic and thermophilic conditions. *Water Science and Technology*, **44(4)**, pp. 97–101, 1997.

[5] Rodríguez, L., Villaseñor, J., Fernández, F.J. & Buendía, I.M., Anaerobic co-digestion of winery wastewater. *Water Science and Technology*, **56(2)**, pp. 49–54, 2007.

[6] Pahl, O., Firth, A., MacLeod, I. & Baird, J., Anaerobic co-digestion of mechanically biologically treated municipal waste with primary sewage

sludge - A feasibility study. *Bioresource Technology*, doi: 10.1016/j.biortech.2007.08.027, 2007.
[7] Gavala, H.N., Yenal, U., Skiadas, I.V., Westermann, P. & Ahring, B.K., Mesophilic and thermophilic anaerobic digestion of primary and secondary sludge. Effect of pre-treatment at elevated temperature. *Water Research*, **37**, pp. 4561–4572, 2003.
[8] Climent, M., Ferrer, I., Baeza, M.d.M., Artola, A., Vázquez, F. & Font, X., Effects of thermal and mechanical pretreatments of secondary sludge on biogas production under thermophilic conditions. *Chemical Engineering Journal*, **133**, pp. 335–342, 2007.
[9] Lin, J., Chang, C. & Chang S., Enhancement of anaerobic digestion of waste activated sludge by alkaline solubilization. *Bioresource Technology*, **62**, pp. 85–90, 1997.
[10] Bolzonella, D., Battistoni, P., Susini, C. & Cecchi, F., Anaerobic codigestion of waste activated sludge and OFMSW: The experiences of Viareggio and Treviso plants (Italy). *Water Science and Technology*, **53(8)**, pp. 203–211, 2006.
[11] Buswell, A.M. & Mueller, H.E., Mechanisms for methane formation. *Industrial & Engineering Chemistry*, **4**, pp. 550–552, 1952.
[12] McCarty, P.L, Anaerobic waste treatment fundamentals. Part one. *Public Work*, **95**, pp. 107–111, 1964.
[13] Maya-Altamita, L., Baun, A., Angelidaki, I. & Schmidt, J.E., Influence of wastewater characteristics on methane potential in food-processing industry wastewaters. *Water Research*, doi: 10.1016/j.watres.2007.11.033, 2008.
[14] APHA-AWWA-WPCF, *Standard methods for the examination of waste and wastewater*. American Public Health Association, Washington, DC, 1998.
[15] R.D. 1310/1990. Real decreto 1310/1990, de 29 de octubre, por el que se regula la utilización de los lodos de depuración en el sector agrario. *Spanish Royal Decree*, 1990.
[16] Zhao, H.W. & Viraraghavan, T., Analysis of the performance of an anaerobic digestion system at the Regina wastewater treatment plant. *Bioresource Technology*, **95**, pp. 815–819, 2004.

The application of an anaerobic baffled reactor to produce biogas from kitchen waste

A. Malakahmad[1], N. Ahmad Basri[2] & S. Md. Zain[2]
[1]*Universiti Teknologi PETRONAS, Malaysia*
[2]*Universiti Kebangsaan Malaysia, Malaysia*

Abstract

The tremendous increase in solid waste generation is an unavoidable occurrence due to the fast growing urbanisation and industrialisation in Malaysia. Anaerobic digestion of organic wastes is receiving more attention in recent years throughout the world because the biomethanogenesis process decomposes organic matter to produce methane gas, which is an excellent energy source as fuel in combined heat and power units. In this study, an application of an anaerobic baffled reactor (ABR) for the production of biogas from kitchen waste was carried out to identify the optimum efficiency of methane gas generation and the potential usage of sludge as organic fertiliser. Different proportions of kitchen waste and activated sewage sludge were mixed and tested in the reactor to achieve the best amount of methane production in the shortest time. Results showed that the combination of 75% of kitchen waste and 25% of activated sewage sludge yielded the best result, which was 74.1% of methane gas. Further, determination for fertiliser value from tests on the sludge in the reactor showed its potential for future use in composting. The amounts of N, P and K were 0.95, 0.80 and 0.45% respectively. According to the observation, anaerobic digestion of kitchen waste in the ABR is able to provide a vital element in an integrated solid waste management and the energy production from this system could be a good reason for many communities to start recycling valuable resources and hence achieving zero waste production.
Keywords: anaerobic baffled reactor, kitchen waste, biogas production.

1 Introduction

In many countries such as Malaysia, it is increasingly more difficult to find suitable locations for landfills, which are accepted by the population. These circumstances are to be found all over the world and make new strategies for waste management necessary. In addition, the promotion of waste minimization and recycling are important components of modern waste management strategies [1].

There are perceptible advantages of anaerobic systems over aerobic metabolic systems and these have been widely reported in the last four decades [2]. There are three clear advantages of the anaerobic treatment over aerobic degradation of organic substrates: the high product and low biomass yield resulting in a limited generation of waste sludge as an unwanted side product; the *in situ* separation of the product as biogas, limiting costs for product separation; and the use of simple technology, as mixing by the biogas produced circumvents the need for other mixing requirements [3].

Currently, digesters are concentrated in developing countries, with over 5 million household digesters constructed in China and India alone [4]. Digesters built around the world vary in their design complexity, construction materials, and costs. In developed countries, digesters often are concrete stirred tank reactors (CSTRs), in which a portion of the produced biogas is utilized to heat the digester [5]. In developing countries, many of the digesters do not have mixing components, do not require continuous monitoring, and are adaptable to any tropical climate [6]. An anaerobic baffled reactor (ABR) initially receives the organic fraction of municipal solid waste (OFMSW) followed by a decomposition process of the materials and eventually produces biogas by microorganisms' activities. Biogas is an excellent source used as fuel in combined heat and power units. The sludge that produced from this system includes body of microorganisms, which can be applied as a source of organic fertilizer [7].

In this study, an anaerobic baffled reactor was proposed for the recovery of energy and production of organic fertilizer. A modified design of ABR was carried out using different combinations of kitchen waste and activated sewage sludge to achieve the highest biogas and methane generation in the shortest time. The sludge produced in the anaerobic system is examined based on its usage as a source of organic fertilizer and to make the ABR a system by completely using all of the materials throughout the system.

2 Methods

2.1 Reactor configuration

Additional vertical baffles in a plug-flow reactor constitute an ABR, which enhances solids retention to allow better substrate accessibility to methanogens [8]. The laboratory-scale unit shown in fig. 1 was made with a total volume of 85 liters. Two tanks as influent tank and effluent tank were designed for feeding and removing the materials to and from the reactor. A gas collector was also provided for collection, calculation and analysis on the amount of biogas. The dimension of the laboratory-scale unit was 75 cm length, 42 cm height and 27 cm depth. The down-flow chambers were 3 cm above the reactor's bottom to route the flow to the center of the up-flow chamber to achieve better contact and greater mixing the feedstock and solids.

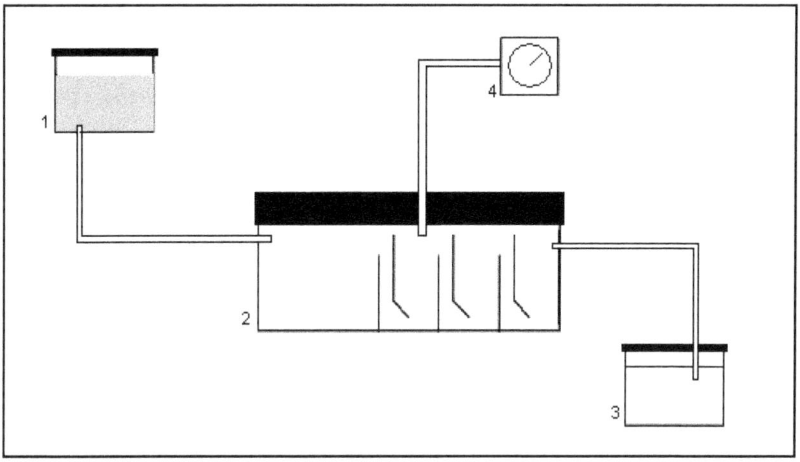

Figure 1: Laboratory scale anaerobic baffled reactor (1: influent tank, 2: ABR system, 3: effluent tank, 4: wet gas meter).

The first compartment of a four-chamber unit, which was 34 liters, was bigger in size while the following compartments were 17 liters. This physical modification provided longer solids retention time and superior performance as compared to the reactor with similar size compartments. The larger compartment in the reactor acted as a natural filter and provided superior solids retention for the small particles. This configuration collected more solid materials than the reactor with four of the similar size chambers [8]. The edges on baffles slanted at 45° to route the flow towards the center of the compartment and, hence, encourage mixing.

2.2 Feedstock preparation

Based on characteristics of kitchen waste and its high C/N ratio [1], different percentages of kitchen waste were mixed with sewage sludge as a source of Nitrogen to normalize the value of C/N. The sewage sludge was collected from a wastewater treatment plant sewage sludge return pipeline and immediately brought to the laboratory. The C/N ratio of sewage sludge used in the experiments, was 11/1 with 5% N of dry weight. Four different combinations of kitchen waste and sewage sludge were compared to determine the best efficiency of methane gas generation in the shortest time (table 1).

Table 1: Percentages of kitchen waste and sewage sludge.

Sample no.	Kitchen waste (%)	Sewage sludge (%)
1	100	0
2	75	25
3	50	50
4	25	75

2.3 Analytical methods

The characteristics of kitchen waste were initially determined. The wastes were taken from kitchen refuse of a cafeteria located at the university. "Part and quarterly methods" were used as standard procedures to prepare samples. Then, they were collected and combined in an approximately equal proportion and mixed thoroughly in the laboratory, shredded and grounded into a size of approximately $1 \times 1 \times 0.5$ cm^3 prior to analysis for chemical composition. Samples were analyzed for moisture content, total solids, total volatile solids, ash content, total organic carbon, Kjeldahl nitrogen, fat, protein, cellulose, hemi cellulose and lignin using analytical methods given by USEPA [9]. The pH of the slurry was measured using a digital pH meter having an accuracy of $\pm$ 0.01 pH unit and the reactor was run in mesophilic conditions. Totally, 40 samples were measured and finally the average of each parameter was calculated. Gas samples were collected by gas sampling injectors and a sample of 100 μl was used for each run. The biogas composition (CH_4 + CO_2) was determined by using a Gas Chromatograph (NUCON 5700) equipped with a thermal conductivity detector and stainless steel column of 6 ft length, 1/4 inch OD, 2 mm ID, containing Porapak Q 100 having mesh range 80–100. The carrier gas used was H_2 and the analysis was carried out at a carrier gas flow rate of 30 ml/min with the injector, column and detector temperatures maintained at 120, 90 and 120°C, respectively. Gas volume was measured using a water displacement method. To measure the fertilizer values, concentration of Nitrogen, Phosphorus and Potassium in produced sludge in the Anaerobic Baffled Reactor (ABR), analytical methods given by APHA [10] were used.

3 Results and discussion

3.1 Kitchen waste characteristics

Table 2 shows the chemical composition of kitchen waste. For the purpose of gas generation the solid content of feed material should be approximately 10–15 percent [11]. Total solids were found between 10.4 to 20.7% with the average of 14.8%. C/N ratio was another value that was calculated. Bacteria normally use up carbon 25–35 times faster than they use nitrogen. Therefore, at this ratio of C/N (25–35/1) the digester is expected to operate at an optimal level for gas production. The C/N ratio for kitchen waste was found 38.2/1.

The substrate consists of complex organic polymers, which have a very important role in the first stage of anaerobic digestion of organic compounds and their presence is vital, because these organic polymers are broken down by extracellular enzymes produced by hydrolytic bacteria, and dissolved in the water. The moisture content of kitchen waste was found 84.5%. The high moisture content verified that kitchen waste was not ideal for incineration or landfill.

Table 2: Chemical composition of kitchen waste.

Parameters	Weight fraction (%) or ratio
Total solids	14.8
Total volatile solids	89.5
Ash	10.5
Total organic carbon	49.7
Kjeldahl nitrogen	1.3
C/N weight ratio	38.2
Fat	8.7
Protein	6.7
Cellulose	14.9
Hemi-cellulose	9.9
Lignin	8.5
Moisture content	84.5

3.2 Volatile fatty acid profile and pH

Changing of pH during anaerobic digestion for all mixtures is illustrated in fig. 2. After 16 days of digestion, the kitchen waste alone was still in its acid formation phase, as indicated by the slightly acidic leachate collected (pH = 5.92). High CO_2 and low CH_4 contents are the best indicators of the digestion process not having reached active methanogenesis [12]. For other samples (samples 2 to 4), the pH variation could be categorized into 3 main zones. The first zone started from the first day till the 4th day, which showed a drastic drop of the pH. This is due to high development rate of volatile fatty acid by the microorganism. The second zone started from the 5th till the 12th day of the experiment. In the second zone, the pH was in the range of 6.9 to 7.3. This is due to the development of CO_3HNH_4 from CO_2 and NH_3, which were produced during the anaerobic process. The percentage of CO_3HNH_4 had caused the increase alkalinity of the samples. Due to this, any differences in the volatile fatty acid content did not affect the pH value. The third zone started on the 13th till the last day of the experiment. In this zone, it was found that the pH value of the samples started to increase. This is due to the development of CO_3HNH_4 still continuing, but no more volatile fatty acid was produced.

3.3 Gas production and composition

Gas production in each sample is illustrated in fig. 3. Kitchen waste alone produced 0.050 m^3/kg VS biogas after 16 days. The combination of 75% kitchen waste 25% sewage sludge had a relatively high gas production rate of 0.594 m^3 per kg VS, while the gas production for samples 3 and 4 was 0.201 and 0.151 m^3/kg VS, respectively.

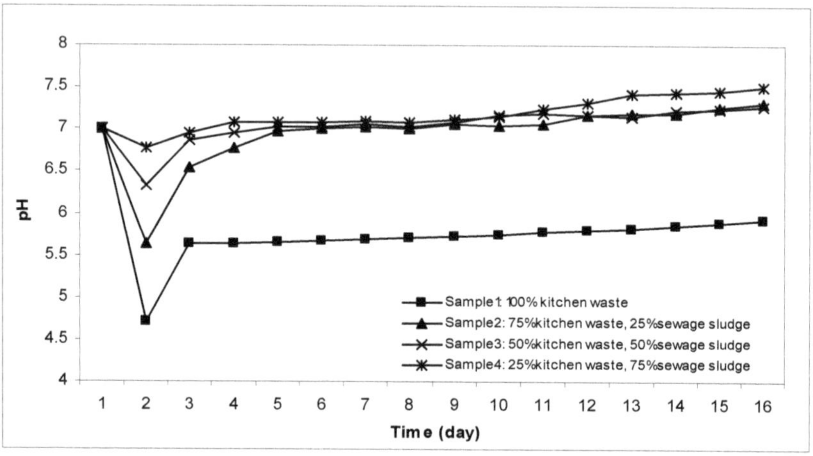

Figure 2: pH variation of the samples during anaerobic process.

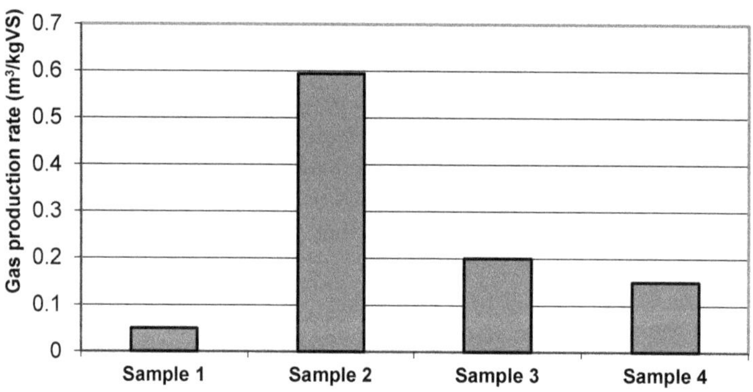

Sample 1: 100% kitchen waste.
Sample 2: 75% kitchen waste, 25% sewage sludge
Sample 3: 50% kitchen waste, 50% sewage sludge
Sample 4: 25% kitchen waste, 75% sewage sludge

Figure 3: Gas production in the mixtures after 16 days.

After 16 days of digestion, the kitchen waste alone was still in its acid formation phase, as indicated by the slightly acidic leachate collected. Furthermore, the relatively high level of CO_2 and low level of CH_4 from kitchen waste alone throughout the whole experimental period (fig. 4) indicated that a longer period was needed for the kitchen waste to go through the acidogenic and acetogenic phases. High CO_2 and low CH_4 contents are the best indicators of the digestion process not having reached active methanogenesis [12].

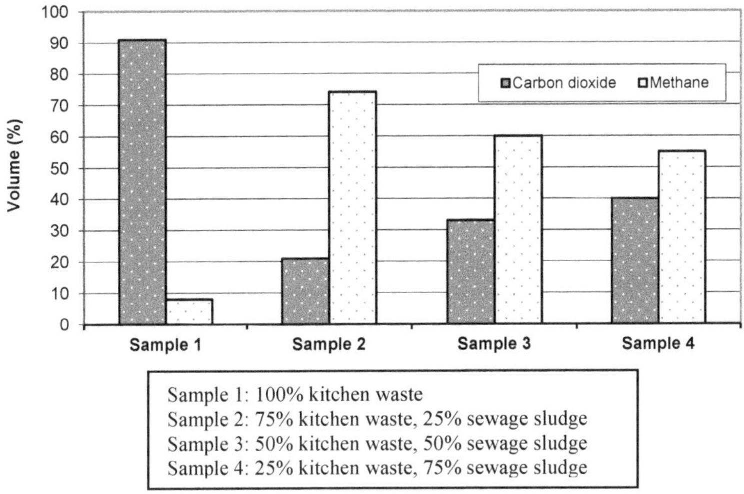

Figure 4: Level of CH_4 and CO_2 in the mixtures after 16 days.

The waste mixtures (samples 2, 3 and 4) produced biogas with high methane content, indicating that the slurries had the proper microbial populations to start the anaerobic degradation. Among them, Sample 2 had the best performance, which may depend on food-to-microorganisms ratio (F/M). The growth of microorganisms in a system is depended on the F/M ratio and if this is decreased, then the microorganisms will go to a death phase and flocs will be formed [13]. From microbiological point of view, the sewage sludge contains a mixture of microorganisms that need food to continue their life and when compared with kitchen waste, sewage sludge has low potential for gas production, as most chemical energy in the sludge has already been depleted in the aeration tank. Polprasert have confirmed that mixing waste at a proper ratio can enhance the digestion process and shorten the time to reach the final phase of anaerobiosis [11]. While gas production for the mixture at 75–25 had started immediately after running the reactor and reached the highest amount of 0.594 kg/m^3 VS after 16 days, the sample with 100% kitchen waste could only produce 0.05 kg/m^3 VS in the same duration (fig. 5).

3.4 NPK values determination

The amounts of N, P and K as the three essential plant microelements were found 0.95, 0.80, and 0.45%, respectively. A comparison between the value of NPK in the ABR and other sources [14] shows that, except from city refuse, the NPK value in produced sludge in ABR is higher than that in rural refuse and also plant residues (fig. 6). However, to compare to chemical fertilizer, the amounts of N, P and K even in city refuse is several time lower and, for better efficiency, the ABR sludge needs to be mixed with chemical fertilizer.

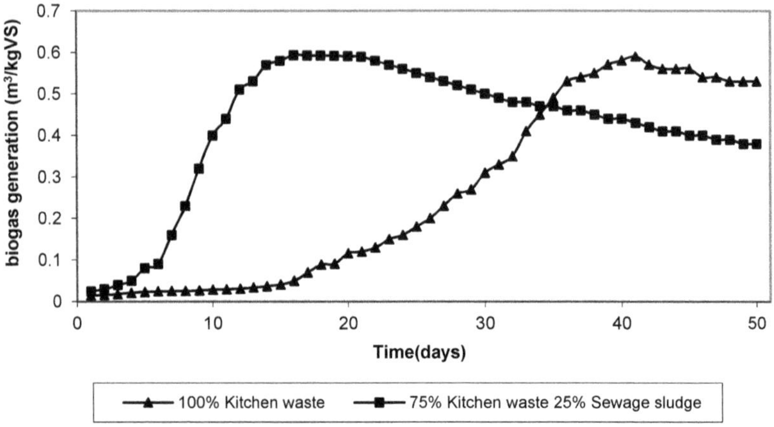

Figure 5: Comparison between 100% kitchen waste and mixture of 75% kitchen waste and 25% sewage sludge.

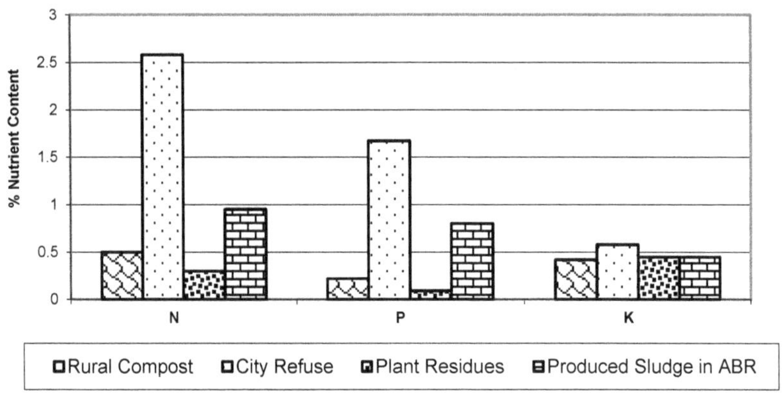

Figure 6: Comparison between NPK values in ABR system with other sources.

4 Conclusion

The environmentally acceptable management of municipal solid waste has become a global challenge due to limited resources, ever increasing population, rapid urbanization and industrialization worldwide. The modified ABR showed the proper performance in biogas production and could be a reasonable choice for energy production as well as a system for the processing of organic parts of produced solid waste. Among all the samples the mixture with 75% of kitchen waste and 25% of sewage sludge had the best biogas and methane production competence due to its suitable C/N and F/M ratio. NPK value in produced sludge in ABR is lower than those present in chemical fertilizers and thus the former needs to be mixed with chemical fertilizer prior to being used as fertilizer.

Acknowledgements

This paper was supported by Universiti Teknologi PETRONAS and Universiti Kebangsaan Malaysia (project code: KK008/2004). Their cooperation with the authors is fully appreciated.

References

[1] Malakahmad, A., Ahmad Basri, N. & Md Zain, S., *Proc. of the 2nd Int. Conf. on Environmental Management*, eds. J.M. Jahi, K. Arifin, S. Surif & S. Idrus, Environmental Management Society Malaysia: Bangi, pp. 41–47, 2004.

[2] Baloch, M.I., Akunna, J.C. & Collier, P.J., The performance of a phase separated granular bed bioreactor treating brewery wastewater. *Bioresource Technology*, **98(9)**, pp. 1849–1855, 2006.

[3] Malakahmad, A., Ahmad Basri, N. & Md Zain, S., *Proc. of the 1st Conf. on Energy and Environment,* eds. K. N. Chee, M. O. Abdullah, A. R. H. Right, M. S. Osman, Universiti Malaysia Sarawak: Kuching, pp. 108–110, 2007.

[4] Huttunen, S. & Lampinen, A., eds., *Bioenergy Technology Evaluation and Potential in Costa Rica*, University of Jyväskylä Printing House: Jyväskylä, Finland, pp. 40–44, 2005.

[5] Berglund, M. & Börjesson, P., Assessment of energy performance in the life-cycle of biogas production. *Biomass and Bioenergy*, **30**, pp. 254–266, 2006.

[6] Lansing, S., Botero, R.B. & Martin, J.F., Waste treatment and biogas quality in small-scale agricultural digesters. *Bioresource Technology*, **99(13)**, pp. 5881–5890, 2008.

[7] Malakahmad, A., Ahmad Basri, N. & Md Zain, S., *Proc. of the LUCED-I&UA Int. Conf. on Environmental Management & Technology,* ed. P. Agumuthu, Malaysian University Consortium for Environment and Development: Putrajaya, pp. 101–109, 2003.

[8] Barber, W.P. & Stuckey, D.C., The use of the anaerobic baffled reactor (ABR) for wastewater treatment: a review. *Water Resource*, **33**, pp. 1559–1578, 1999.

[9] Test Methods for Evaluating Solid Waste, Physical/ Chemical Methods; U.S. Environmental Protection Agency, Office of Solid Wastes, SW-846 Online. www.epa.gov/epaoswer/hazwaste/test/main.htm

[10] APHA – AWWA, *Standard Methods for Water and Wastewater Examinations*. 21st ed. American Public Health Association/American Water Works Association: Washington, DC, 2005.

[11] Polprasert, C., *Organic Waste Recycling*, 2nd ed., John Wiley and Sons Publication: Chichester, pp. 25–47, 1996.

[12] Pohland, F.G. & Ghosh, S., *Developments in Anaerobic Treatment Processes in Biological Waste Treatment*, 2nd ed., Canale Interscience Publication: New York, 1971.

[13] Shariat, M., *Principle of Microbiology in Water and Wastewater Treatment*, Publication of Tehran University: Tehran, 2002.

[14] Polprasert, C., Edwards, P., Pacharaprakiti, C., Rajpot, V.S. & Suthirawut, S., *Recycling rural and urban nightsoil in Thailand*, AIT Research Report, Asian Institute of Technology: Bangkok, 1982.

Rubbish or resources: an investigation into converting municipal solid waste to bio-ethanol production

A. Li & M. Khraisheh
University College London, UK

Abstract

An investigation into the possibility of replacing conventional biomass with biodegradable municipal solid waste (BMSW), which provides an alternative solution for preventing the organic fraction of municipal solid waste going into landfill as required by EU Landfill Directives, was carried out. The experimental results have shown that more than 85% of the cellulose from the waste can be converted to glucose, which can be easily fermented to ethanol production. The conversion rate is relatively high compared with other conventional biomass. Taking into account the existing waste collection system and the cost for disposing of waste makes this waste-to-ethanol system economically valuable. However, every type of energy has some advantages as well as disadvantages. This paper identifies the potential impacts of using BMSW as bio-ethanol resource from the point of view of both energy development and waste management. The opportunities of MSW-to-ethanol as an alternative waste management are also discussed with the recommendation for further life cycle study.
Keywords: biodegradable municipal solid waste (BMSW), MSW-bioethanol, environmental impact, life cycle assessment.

1 Introduction

The use of bio-fuel for transport is becoming increasingly important for a number of reasons, such as environmental concerns relating to climate change, depletion of fossil fuel reserves and reduction of reliance on imports. On the other hand, European legislative pressures [1] target for minimizing landfill use: the amount of biodegradable municipal solid waste (BMSW) going to landfill must be reduced by 25% by 2010, 50% by 2013 and 65% by 2010. Thus, an alternative way of disposing of BMSW needs to be investigated.

London Boroughs (UK) collected 4.4 thousand tones of total municipal solid waste (MSW) in 2004–2005 [2]. Among the waste generated, about 73% goes to landfill and 17% to incineration. This study investigates the possibility of replacing the conventional biomass with biodegradable municipal solid waste, which provides an alternative solution for preventing the organic fraction of municipal solid waste going into landfill and producing bio-ethanol production for the renewable energy market demand and discusses the potential environmental impacts of this process.

The process of converting BMSW to ethanol is shown in fig. 1. Pre-hydrolysis treatment and enzymatic hydrolysis of BMSW are of crucial importance during the bioconversion of BMSW-to-bio-ethanol, and thus their optimization will result in beneficial environmental and economic practices [3]. A previous study has investigated the impact of pre-hydrolysis treatments (i.e., dilute acid type and concentration, steam treatment, microwave) and enzymatic hydrolysis (i.e., enzyme type and concentration) on the bioconversion of BMSW-to-glucose at laboratory-scale [4]. This study is to continue the previous research and optimize the processes by using mixed municipal solid waste for bio-ethanol production and bio-ethanol technology.

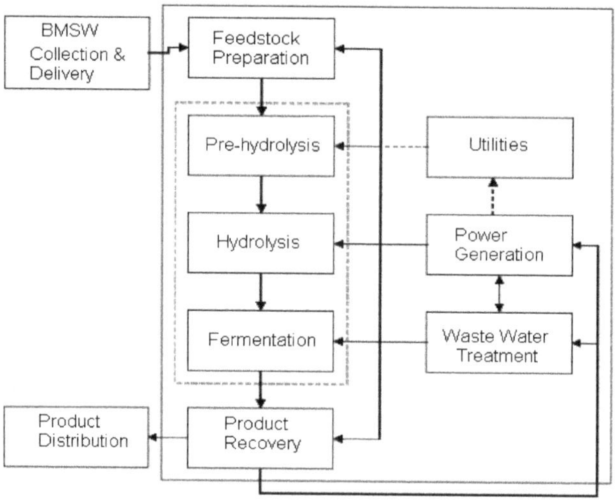

Figure 1: BMSW-to-bioethanol process.

2 Materials and methods

The BMSW-to-ethanol process is shown in fig. 1 which includes sample collection, pre-treatment, hydrolysis, fermentation, product recovery and product distribution. The BMSW used in this study was selected from typical biodegradable waste, such as kitchen waste, garden waste and paper waste. The sample consists of 20% carrot peelings, 20% potato peelings, 20% grass, 20% newspaper, and 20% scrap paper. The selected waste was milled to small particles with size of 0.2–1.2 mm, and then followed by pre-hydrolysis with sulphuric acid and steam treatment, and then enzymatic hydrolysis.

2.1 Pre-hydrolysis treatment

Biomass and dilute sulphuric acid (1%) were mixed at a solid/liquid ratio of 0.5 g in 50 ml. The acid-soaked samples were then placed in an autoclave at a constant temperature 121°C for 60 minutes. Following steam treatment, the samples were washed with deionised water until the pH of the filtrate remained constant at an approximate value of 4.5. The pre-treated samples were drained and air-dried overnight to a final 45% moisture content.

2.2 Enzymatic hydrolysis

Enzymatic hydrolysis of pre-hydrolysed substrate was carried out within the study temperature range in vials (50 ml) placed in an orbital agitator at 68 rpm for 24 h. The liquor pH was adjusted using 0.1 M citrate acid-sodium citrate buffer. Enzymes, T. viride (Sigma, UK) were used as received in combination with the enzyme beta-glucosidase (Sigma, UK). The cellulase activity of the enzymes [5] was measured by the Filter Paper assay developed by Mandels *et al.* [6], and the activity was expressed in terms of filter paper units (FPU). The b-glucosidase activity was measured by the PNPG assay [7], and reported as PNPG units (PNPGU). Samples were withdrawn from the reaction media after 24 h.

2.3 Analytical methods

Sugar concentrations were determined routinely from centrifuged samples (13,000 rpm, 5 min) using HPLC (Perkin-Elmer Ltd, UK). All compositional analyses were performed in triplicate and data were corrected to a 100% dry matter (DM) basis. Glucose yield as a percentage of the theoretical yield [percentage digestibility, obtained from the equation which involves the transfer of cellulose to sugar $(C_6H_{10}O_5)_n + nH_2O = (C_6H_{12}O_6)n$] was computed using the formula given by the National Renewable Energy Laboratory (Standard Biomass Analytical Procedures) [8]. A factorial experimental design with the software package Design Ease 6.0.6 (Stateease, Inc.) was used in this study. Selected factors identified as more convenient to control the hydrolysis process were selected from enzymatic hydrolysis. The relevant factors with the study range are shown in table 1.

The mass balances were calculated in two main steps: (1) from raw material to pre-hydrolysed biomass, and (2) to enzymatic hydrolysis residues, with consideration of the weight loss during each step.

The environmental impact assessment is based on the results from laboratory work. The assessment covers a number of environmental issues such as waste management, carbon dioxide emission, water issue, land-use and biodiversity, by comparing with traditional biomass such as corn and energy crops.

Table 1: Study factors for the enzymatic hydrolysis process.

Factor	Study range
A: Particle size	2–20 mm
B: Substrate concentration	5–15 g/l
C: Enzyme loading	10–100 FPU
D: Beta-glucosidase	10–100 PNPGU g-1
E: pH	3.7–5.0
F: Temperature	37–50°C

3 Experimental results and discussion

A previous study [4] showed that BMSW as a feedstock produced about 52% glucose for the production of ethanol. Recent studies on hydrolysis optimization showed about 85% glucose yield can be reached under the optimal conditions (lower substrate concentration, lower beta-glucosidase loading and temperature 50°C) (fig. 2).

The contribution of substrate concentration to the glucose yield was significantly higher (93.6%), than the contribution of temperature (3.2%) and beta-glucosidase loading (2.6%). Other factors such as particle size, enzyme loading and pH have very little effect on the glucose yield within the study range. The results obtained by the model presented a satisfactory correlation to the actual values (R2 = 0.90). Additionally, this study indicated that lower substrate concentration (5 g/l), higher temperature (50°C) and less beta-glucosidase produced the highest glucose yields. The significant contribution of substrate concentration and temperature at 50°C has been in agreement with other studies on conventional biomass [9]. The reason for less beta-glucosidase may be concluded as the inhibition of the enzyme as suggested by Gregg and Saddler [10]. This study indicated that BMSW fractions did not require higher cellulase loadings to obtain high glucose yields, which can reduce the cost of the process.

Results from the mass balance revealed that about 0.25 g biomass remained after pre-hydrolysis and 0.1 g remained after the enzymatic hydrolysis process when 0.5 g BMSW was used as feedstock. This indicated that about 80% of the biodegradable waste fractions were removed by using this BMSW-to-ethanol process.

4 Environmental impacts

The environmental impacts were assessed from the following areas: waste management, climate change, water issues, land use, biodiversity and others.

(a)

Factor	% Contribution
B - Substrate Concentration	93.6
D - Beta-glucosidase	2.6
F - Temperature	3.2

(b)

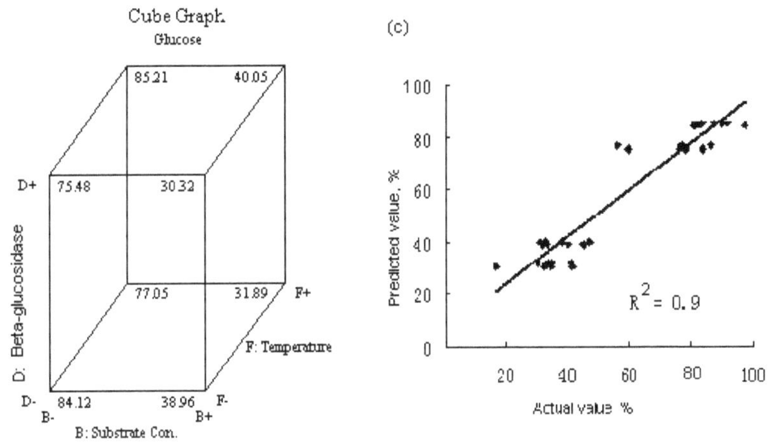

Figure 2: Factorial experimental design applied to BMSW fraction. (a) Factor effects on glucose yield, (b) cube graph of glucose yield, (c) comparison of actual value and predicted value.

4.1 Waste management

Using BMSW as resource for bio-ethanol production will reduce the biodegradable fraction from MSW going to landfill which is one of the requirements of EU Landfill Directives. According to the mass balance from the laboratory plant, the amount of waste will be massively reduced by 80% and the main composition of solid residues from the hydrolysis process are acid insoluble lignin, ash content, protein absorbed and dead cells, which can be used as compost. This technology can be used as an alternative solution. However, sustainable waste management needs to involve different types of waste management methods in order to minimize the waste produced, maximize energy/material recovery and meet the needs of environmentally, economically and socially sustainable.

4.2 Climate change

Ethanol represents the closed carbon dioxide cycle because after burning of ethanol, the released carbon dioxide is recycled back into plant material since plants use carbon dioxide to synthesize cellulose during photosynthesis cycle

[11, 12]. Ethanol production process only uses energy from renewable energy sources; no net carbon dioxide is added to the atmosphere, making ethanol an environmentally beneficial energy source. In addition, the toxicity of the exhaust emissions from ethanol is lower than that of petroleum sources [13]. Apart from these, using BMSW as ethanol resources will avoid the biodegradable fraction going to traditional waste disposal plant such as landfill or incineration, which often causes green gas emissions.

4.3 Water issues

A large quantity of water is required for growing energy crops, which causes water shortage problems but there are also cases where there is plentiful water. One of the advantages is that irrigation can increase productivity surrounding plantations. For the ethanol process itself from both BMSW and energy crops, to produce 1 litre ethanol, 4–8 litres of water are required [14]. Moreover, 13 litres of sewage effluent are produced per 1 litre of ethanol produced because of the water that has to be added to the ground corn grain for the fermentation process [15]. Water remains a big issue for bio-ethanol process.

4.4 Land use

Growing energy crops for bio-ethanol purposes requires lots of land use. Harvesting corn stover may result in lowering soil organic carbon levels and soil nitrogen content, and may also increase soil erosion [16]. If agricultural prices are increasingly linked to energy prices then impacts on food security will also mean more volatility in the markets. World prices usually account for less than 25% for any variation at country level. It is currently hard to model the effect of bio-fuels [17]. Using BMSW as biomass can free the land used for landfill and growing energy crops.

4.5 Biodiversity

Input intensive energy crop production involves the movement of agrichemicals, especially nitrogen, phosphorus, and pesticides from farms to other habitats and aquifers. The use of BMSW for bio-ethanol purpose can avoid this problem.

Apart from the environmental impacts discussed above, this BMSW-to-ethanol system has its economic and social benefits. The end product: bio-ethanol is a valuable market product, which brings economic benefits from the non-valuable waste and meets the requirement of current energy market demands. However, this process requires the biodegradable waste fraction to be separated from non-biodegradable wastes from the households. The economic benefits with bio-fuel product will encourage waste separation.

5 Opportunities of MSW-to-ethanol as an alternative waste management

Waste management varies from developing nations to developed countries. Under different economic and social environment, different types of waste management methods from re-use, recycle and energy/material recovery to disposal are used. Taking the waste management system in London as an example, if all of London's BMSW available for new recovery options were used for bioethanol production, an estimated 346 million litres of ethanol could be produced by 2020, providing 14.6% of London's 2004 petrol car consumption. This evaluation should however be considered rough and further consideration should to be given to a wider range of performance criteria and system requirements.

To realise the potential for bioethanol production requires performance across the lifecycle of the fuel chain against other fuels to be competitive within the end-product (biofuel and wider transportation fuel) market and the feedstock (waste resource) market. Net energy balance, greenhouse gas balance and production costs are fundamental indicators for sustainable performance. There is a lack of published research examining this in regards to MSW-bioethanol by biochemical processes. Three important areas of research require attention:

1. Lifecycle analysis of each of the competing MSW-ethanol technologies converting different BMSW feedstocks;
2. A biofuel lifecycle comparative study which includes the optimum current and future MSW-bioethanol technologies; and
3. A waste management lifecycle comparative study which includes the optimum current and future MSW-bioethanol technologies.

References

[1] European Commission (EC). Directive 2003/30/EC of the European Parliament and of the Council of 8 May 2003 on the promotion of the use of biofuels or other renewable fuels for transport. *Official Journal of the European Union L*, **123**, pp. 42–46, 2003.

[2] Defra (Department for Environment, Food and Rural Affairs). Municipal Waste Management. 2006. http://www.defra. gov.uk/environment/statistics/ wastats/index.htm

[3] Mosier, N., Wyman, C., Dale, B., Elander, R., Lee, Y.Y., Holtzapple, M. & Ladisch, M., Features of promising technologies for pretreatment of lignocellulosic biomass. *Bioresource Technology*, **96**, pp. 673–686, 2005.

[4] Li, A., Antizar-Ladislao, B. & Khraisheh, M., Bioconversion of municipal solid waste to glucose for bio-ethanol production. *Bioprocess and Biosystems Engineering*, **30(3)**, pp. 189–196, 2007.

[5] Melander, C., Adden, R., Brinkmalm, G., Gorton, L. & Mischnick, P., New approaches to the analysis of enzymatically hydrolyzed methyl cellulose. Part 2. Comparison of various enzyme preparations. *Biomacromolecules*, **7**, pp. 1410–1421, 2006.

[6] Mandels, M., Andreotti, R. & Roche, C., Measurement of saccharifying cellulose. *Biotechnology and Bioengineering Symposium*, **6**, pp. 21–33, 1976.
[7] Paquot, M. & Thonart, P., Enzymatical hydrolysis of regenerated cellulose. *Holzforschung*, **36**, pp. 177–181, 1982.
[8] NREL (National Renewable Energy Laboratory), SSF Experimental protocols: Lignocellulosic Biomass Hydrolysis and Fermentation, 1995. http://www.devafdc.nrel.gov/pdfs/4691.pdf
[9] Ballesteros, M., Oliva, J.M., Manzanares, P., Negro, M.J. & Ballesteros, I., Ethanol production from paper material using a simultaneous saccharification and fermentation system in a fed-batch basis. *World Journal of Microbiology and Biotechnology*, **18(6)**, pp. 559–561, 2002.
[10] Gregg, D.J. & Saddler, J.N., Factors affecting cellulose hydrolysis and the potential of enzyme recycle to enhance the efficiency of an integrated wood to ethanol process. *Biotechnology and Bioengineering*, **51(4)**, pp. 375–383, 1996.
[11] Wyman, C.E., Biomass ethanol: technical progress, opportunities, and commercial challenges. *Annual Review of Energy and the Environment*, **24**, pp. 189-226, 1999.
[12] Chandel, A.K., Chan, E.S., Rudravaram, R., Narasu, M.L., Rao, L.V. & Ravindra, P., Economics and environmental impact of bioethanol production technologies: an appraisal. *Biotechnology and Molecular Biology Review*, **2(1)**, pp. 014–032, 2007.
[13] Wyman, C.E. & Hinman, N.D., Fundamentals of production from renewable feedstocks and use as transportation fuel. *Applied Biochemistry and Biotechnology*, **24/25**, pp. 735–775, 1990.
[14] Lang, K., *Shellulosics – The Next Generation?* Canada, 2007.
[15] Geotimes, Weighing in on Renewable Energy Efficiency, August 2005.
[16] Mann, L., Tolbert, V. & Cushman, J., Potential environmental effects of corn (Zea mays L.) stover removal with emphasis on soil organic matter and erosion. *Agriculture, Ecosystems & Environment*, **89(3)**, pp. 149–166, 2002.
[17] Overseas Development Institute (ODI), Biofuels and poverty reduction: Is there a way through the maize? London, UK. 27th June, 2007.

Biohydrogen production by *Clostridium beijerinckii*

M. T. Chmiel & V. Yargeau
Department of Chemical Engineering, McGill University, Canada

Abstract

Climate change, along with the rapid depletion of petroleum and natural gas reserves, has prompted many to search for renewable and environmentally friendly energy options. Hydrogen has been identified as a possible alternative to fossil fuel energy. Biological hydrogen production from organic substrates can be achieved by a two-stage approach combining the anaerobic and photosynthetic continuous processes in series. As a first step, an investigation of the production of hydrogen from glucose by *Clostridium beijerinckii* was conducted. A study examining the effect of initial pH (range 5.7 to 6.5) and COD loading (range 1 to 3 g/L) on the specific conversion and specific hydrogen production rate has shown interaction behaviour between the two independent variables. The highest conversion of 10.3 mL H_2/(g COD/L) was achieved at pH of 6.1 and COD of 3 g/L, whereas the highest production rate of 71 mL H_2/(h*L) was measured at pH 6.3 and substrate loading of 2.5 g COD/L. In general, there appears to be a strong trend of increasing hydrogen production rate with an increase in both substrate concentration and pH. The current work focuses on the identification of soluble metabolites such as ethanol, acetate and butyrate in order to evaluate the possible shift in metabolism resulting from varying initial conditions and micro-nutrients availability. The results obtained so far, along with some preliminary experiments using industrial wastewater as substrate, indicate the possible use of such waste streams for the production of biohydrogen.
Keywords: biohydrogen, wastewater, clostridium.

1 Introduction

Since the industrial revolution, humans have been heavily dependent on fossil fuels for their energy source and, with the growing economy and globalization, that dependence is only expected to intensify. In 2003, the Energy Information Administration (EIA) estimated the total world energy consumption at 420.7 quadrillion BTU, a number which is projected to increase to almost 510 quadrillion BTU by the year 2010 [1]. This continuing demand has already put an enormous strain on the available fossil fuel reserves and there is growing concern about energy supply and security. The 2003 NEB report was predicated on an assumption that fossil fuel prices would remain stable at US$22 per barrel WI (West Texas Intermediate) until 2025; in reality, WI prices for 2006–2008 were expected to triple to an average of US$65 per barrel with peaks as high as US$80 [1]. Peaks as high as US$96 were observed in 2007, reflecting a dramatic rise in price and uncertainty in market supply.

In the 2005 World Energy Report, the German Federal Institute for Geosciences and Natural Resources (BGR) estimated that more than 60% of the total potential petroleum has been extracted in North America and almost 50% in Western Europe [2]. Similarly, the total global natural gas (NG) reserves were reported at 6112 trillion cubic feet as of January 1, 2006, and the 2003 global annual NG consumption was 95 trillion cubic feet [1]; without accounting for the growth in demand, this would suggest that our resources will run out in approximately 60 years. Despite the claim by some that evolving technology will allow us to tap into currently unavailable energy sources, our present reliance on finite energy resources cannot be sustainable in the long term. Such over reliance on a very narrow portfolio of energy supply stocks has prompted serious interest in the development of sustainably produced replacement fuels.

The increasing energy consumption has not only put a strain on our raw resources, but also on our planet and its ecosystems. Incomplete combustion processes, which result from the burning of fossil fuels, produce a great amount of gases such as carbon dioxide (CO_2) and nitrogen oxides (NOx). Carbon dioxide emissions are of particular concern, since CO_2 has been identified as a greenhouse gas (GHG). The atmospheric concentration of GHGs has been steadily rising and in 2005 the concentration of CO_2 rose to a record high of 378.9 ppm [3]. This increase has been directly linked to human activity; over 25,000 million metric tons of CO_2 were emitted in 2003 [1], a value that is projected to grow over the years. High concentrations of GHGs have resulted in a rise of the average surface temperatures as well as adverse effects on weather patterns, human and animal life.

Emission of greenhouse gases, along with the rapid depletion of oil and gas reserves, has prompted many to search for renewable and environmentally friendly energy alternatives. Among others, hydrogen has been identified as a possible alternative to fossil fuel energy and a worldwide investigation of hydrogen as a future energy carrier is now underway. The current hydrogen production methods can be grouped into three main categories: fossil fuel processing which includes natural gas, coal, and biomass as the process input; electrolysis of water using conventional, renewable, or nuclear energy as the electricity source; and modern approaches which include photo-synthetic and photo-biological splitting of water [4, 5]. Hydrogen is widely recognized as an environmentally friendly energy carrier and, when generated from renewable residues, promises a substantial reduction in the overall GHG emissions.

Biological production of hydrogen is proposed as a promising avenue for the production of hydrogen because it is less energy intensive, more environmentally friendly and can be based on various substrates. Biological hydrogen production processes can be classified as follows: 1. Biophotolysis of water using algae and cyanobacteria; 2. Photodecomposition of organic compounds by photosynthetic bacteria; 3. Fermentative hydrogen production from organic compounds, and 4. Hybrid systems using photosynthetic and fermentative bacteria. Among these, the photosynthetic bacteria are favoured candidates for large-scale production due to their high substrate conversion efficiencies and their capability of producing hydrogen from a wide variety of substrates [6]. Fermentative bacteria are also favourable candidates due the lower operating costs (when compared to photosynthetic bacteria), because there are no light requirements and growth of the microbes is simpler. Although the technical feasibility of fermentative hydrogen production from a variety of substrates has been well demonstrated in literature, the technology is still at its infancy [7]. Reviews [8, 9] demonstrate that most research work has been carried out using batch processes, while continuous set-ups are potentially more attractive. Previous studies on the production of biohydrogen using both fermentative and photosynthetic bacteria have also shown that research is still needed to improve the conversion efficiencies [10] and the stability of the hydrogen production [11].

Due to inherit thermodynamic limitations associated with the fermentative approach, a two-step biological hydrogen production process has been suggested. By combining the anaerobic and photosynthetic steps, higher overall substrate conversion efficiency is possible, since the photosynthetic microbes can degrade soluble metabolites from the fermentative step using sunlight to overcome the energy barrier. Recently, Kim *et al.* [12] achieved conversion efficiency 30% higher than previously reported, using a two-step process which utilized *Clostridium butyricum* and *Rhodobacter sphaeroides* in the fermentative and photosynthetic degradation, respectively, from a *C. reinhardtii* biomass (mainly glucose-starch) substrate.

A long list of possible feedstocks could be considered as potential substrate for biohydrogen production. The use of various waste streams such as glycerol containing waste from the biodiesel production, non-woody residues and agricultural products seems to be an interesting avenue. However, further improvement of the technology and conversion efficiency is still required. This paper aims to investigate the feasibility of using industrial waste water as a substrate for biological hydrogen production using a pure culture bacterium of *Clostridium beijerinckii*, with hopes of creating an economically attractive process in the long term.

2 Materials and methods

2.1 Bacterial cultivation

Pure culture of *Clostridium beijerinckii* was purchased from ATCC (#8260) and used for the duration of the study. The bacterium was pre-cultured in the recommended nutrient broth (Difco™ Reinforced Clostridial Medium) at 30°C in serum bottles inside a dark incubator shaker. The bacteria were transferred 2–3 times and cultivated for 12 hours (till stationary phase) between each transfer prior to use in the experiments in order to ensure a healthy and active culture population.

2.2 Experimental procedure

Batch hydrogen production experiments were carried out in 100 mL serum bottles with a working volume of 50 mL. The media contained a prescribed amount of glucose as well as essential growth nutrients [13] (in mg/L) 850, KH_2PO_4; 750, K_2HPO_4; 3, H_3BO_3; 200, $MgSO_4.7H_2O$; 1, $Na_2MoO_4.2H_2O$; 1, $ZnSO_4.7H_2O$; 2, $MnSO_4.4H_2O$; 0.1, $Cu(NO_3)_2.3H_2O$; 1, $CaCl_2.2H_2O$; 2, EDTA; 12, $FeSO_4.7H_2O$; 4, thiamine; 3, biotin; 5, *p*-aminobenzoic acid; 6.5, nicotinamide; 420, glutamic acid; 1, resazurin. A 0.1 M phosphate buffer was also added, in order to prevent pH decrease due to organic acid accumulation during the bacterial metabolism. The initial pH was adjusted to a desired value using 5N NaOH or 5N HCl. The media was dispensed into the bottles, which were then flushed with oxygen-free argon gas, capped with a butyl rubber stopper, sealed with an aluminium crimp and sterilized in an autoclave. In order to remove any residual oxygen in the media, prior to inoculation a reducing agent, $Na_2S.9H_2O$, was added at a 0.025% (w/w) concentration. All bottles were inoculated with 3% (v/v) *C. beijerinckii* in the stationary phase and incubated in an orbital shaker at 30°C $\pm$ 1°C and rotational speed of 180 rpm. In total, thirteen experiments were conducted each in triplicate to ensure reproducibility.

2.3 Factorial design

In order to analyze the effect of initial glucose concentration and initial pH as well as any interactions between the two variables, a fractional factorial design was employed [14]. A nine trial design was constructed as shown in fig. 1 to

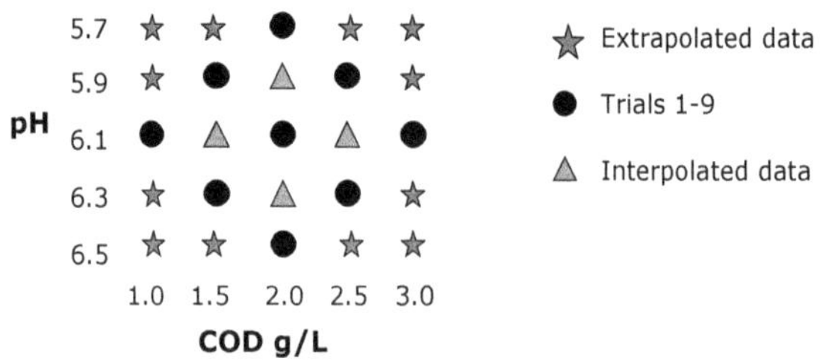

Figure 1: Fractional factorial design.

cover the area of interest. The substrate concentration varied from 1 to 3 g COD/L with the central value of 2 g COD/L and the pH varied from 5.7 to 6.5 with a central value of 6.1. Both ranges were based on values previously observed in a number of local industrial wastewater streams (data not shown). The pH range was modified from the original design of 4.5–6.5 after no growth was observed in the lower end of the range.

2.4 Analytical methods

Biogas production was periodically measured using 50, 25 and 10 mL glass syringes fitted with hypodermic needles as described by Owen *et al.* [15]. The biogas was sampled from the headspace with a 2 mL gas tight syringe and analyzed for the amount of hydrogen (H_2%) using a gas chromatogram (Hewlett-Packard 5890) equipped with a thermal conductivity detector (TCD). The GC was fitted with a stainless steel molecular sieve column (6" x 1/8") and the injector, oven and detector temperatures were set at 100°C, 80°C and 100°C, respectively. Argon was used as the carrier gas with a flow rate of 2 mL/min. Final glucose concentrations were analyzed using a glucose (HK) assay kit (Sigma-Aldrich).

2.5 Data analysis

The cumulative hydrogen gas production curves were constructed as previously described [16] by measuring the gas composition in the headspace of the bottle and the total volume of the biogas produced at each time interval and applying the mass balance equation

$$V_{H,i} = V_{H,i-1} + C_{H,i}(V_{G,i} - V_{G,i-1}) + V_H(C_{H,i} - C_{H,i-1}) \quad (1)$$

where $V_{H,i}$ and $V_{H,i-1}$ are cumulative hydrogen gas volumes at the current *(i)* and previous (i–1) time interval, $V_{G,i}$ and $V_{G,i-1}$ are the total biogas volumes at the current and previous time interval, $C_{H,i}$ and $C_{H,i-1}$ are the fractions of hydrogen gas in the headspace of the bottle as determined by gas chromatography in the current and previous interval and V_H is the total volume of headspace in the bottle.

Each of the cumulative hydrogen production curves was modeled using the modified Gompertz equation

$$H(t) = H_{max} \exp\left\{-\exp\left[\frac{R \times e}{H_{max}}(\lambda - 1) + 1\right]\right\} \quad (2)$$

where $H(t)$ is the cumulative hydrogen production (mL) during the course of the incubation time, t (hours), H_{max} is the hydrogen production potential (mL), R is the maximum production rate (mL H_2/h) and λ is the duration of the lag phase (h). This model has been commonly used to describe biological production of various gases such as methane, hydrogen and biogas in a batch set up [17–19]. The cumulative production curves were fit using Matlab 6.5 by minimizing the sum of square error (SSE). Initial estimates for the parameters (H_{max}, R and λ) were selected based on visual inspection. The hydrogen production potential was normalized with respect to substrate concentration to give the conversion efficiency, P_s (mL H_2/g COD/L); the specific H_2 production rate, R_s was defined as R/V_{media} (mL H_2/h-L).

3 Results and discussion

In order to assess the effect of substrate concentration and initial pH on the conversion efficiency and the specific production rate of hydrogen, the data obtained from the Gompertz modelling was graphed in two three-dimensional plots. A second-order polynomial regression was conducted in order to interpolate/extrapolate results to cover the entire region of interest as outlined in fig. 1. The main objective of the regression was to enhance visual understanding of the types of trends that exist within the matrix.

As shall be shown in Section 3.3, the extrapolated regions turned out to be of particular interest upon visual inspection. Four additional experiments, covering the corner points of the matrix, were carried out in order to validate these initial results. The revised fractional factorial design can be seen in fig. 2.

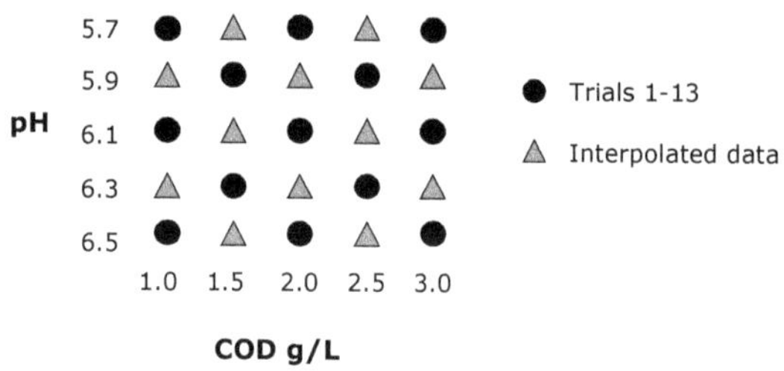

Figure 2: Revised fractional factorial design.

By eliminating extrapolated regions from the design, a more concrete conclusion can be reached with regards to the location of the area where initial pH and substrate concentration provide the most promising operational conditions.

3.1 Specific hydrogen production potential

The specific hydrogen production potential, or the conversion efficiency, fitted based on the original design, is plotted in fig. 3a. It is clear from the data that COD loading and initial pH play a role on the yield. The trends indicate that higher conversion is achieved at higher glucose loading and a mid-range pH in the vicinity of 6.0; the highest P_s of 10.3 mL H_2/g (COD/L) was measured at substrate concentration of 3 g COD/L and pH of 6.1. Both high and low-end pH appears to be unfavourable, but this effect is secondary to that of substrate concentration where the difference between the highest and lowest expected value is almost double.

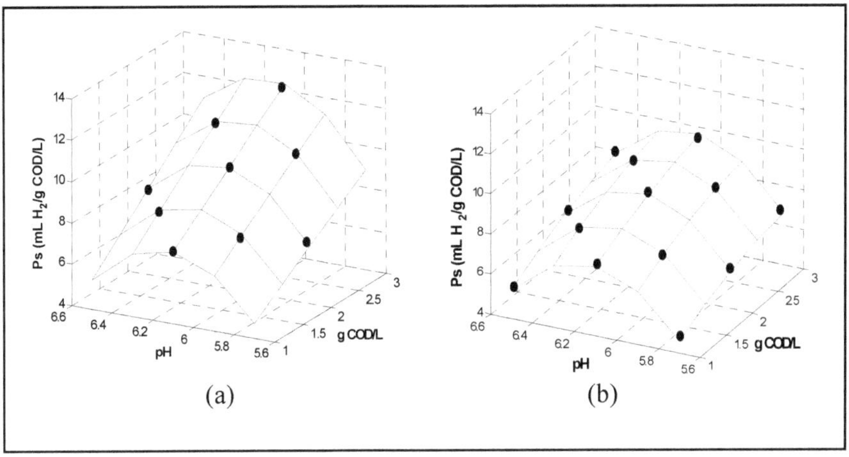

Figure 3: Specific hydrogen production potential (a) based on the fractional factorial design (9 trials); (b) based on the revisited design (13 trials). The marked points indicate the sum of measured values and the residual of the regression.

When the extra four trials (corner points) were added to the matrix, however, changes were observed in the regression of the response variable, P_s (see fig. 3b). In general, the trends discussed above remain intact, though their degree does not appear to be as pronounced. Specifically, increasing substrate concentration appears to play a smaller role on the yield than initially anticipated.

3.2 Specific hydrogen production rate

The specific hydrogen production rate obtained from the nine-trial data is shown in fig. 4a. The data indicates that both initial pH and glucose concentration have noticeable effects on the rate of hydrogen production and also that there is interaction between these two parameters. At low substrate loadings, the largest R_s appears to be located in the middle of the pH range, in the vicinity of 6.1. As the glucose concentration increases, however, the high-end pH values seem to favour greater hydrogen production rates. Low pH gives poor R_s across the entire substrate concentration span. This is somewhat expected, as similar trends have been reported in other works [17, 19]; low pH tends to have an inhibitory effect on the bacteria causing a longer lag phase and lower rate of production.

Based on fig. 4a, the largest R_s appears to be at high initial pH and substrate concentration, a region of the plot that is mostly extrapolated. As a result, conducting experiments at the corners of the matrix became crucial prior to drawing final conclusions regarding the location of highest hydrogen production rate. The new surface plot can be seen in fig. 4b. In general the shape, as well as the scale of the graph, remained unchanged, with slight alterations to the

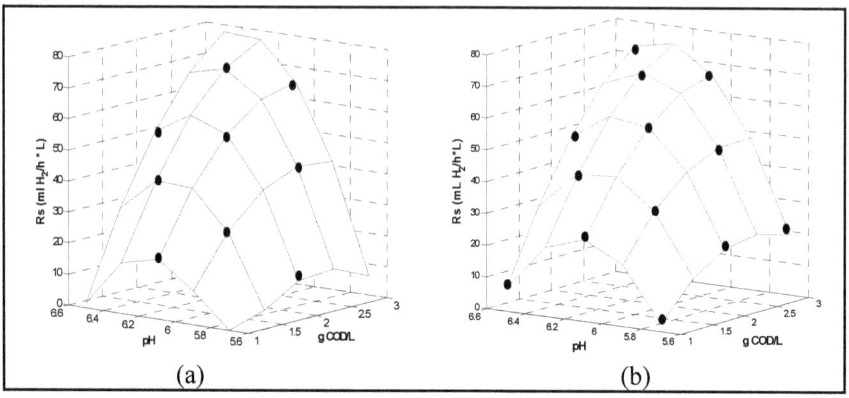

Figure 4: Specific hydrogen production rate (a) based on the initial fractional factorial design (9 trials); (b) based on the revised design (13 trials). The marked points indicate the sum of measured values and the residual of the regression.

extrapolated areas. The highest R_s of 71 mL H_2/(h*L) was measured at pH 6.3 and substrate loading of 2.5 g COD/L, and the region around these parameter values seems to consistently produce similarly high rates.

In practice, any treatment/conversion process strives for high yield and fast rate but often, such as in this case, the most promising operating conditions are different for each variable of interest. When this occurs, a compromise must be made. In this system, out of the two response variables, P_s and R_s, the rate is of more interest since continuous biogas systems operate more optimally at low to mid-range hydraulic retention times [20, 21] meaning that maximum conversion is hardly ever achieved. For this reason, more effort should be devoted to maximizing the rate of hydrogen production with partial sacrifice to the conversion efficiency.

4 Future work

Current work is focusing on the identification and quantification of the soluble metabolites from the batch experiments such as ethanol, acetate and butyrate in order to evaluate the possible shift in metabolism with varying operating conditions. Future work will involve characterization of industrial wastewater samples, testing their potential as a substrate for biohydrogen production, as well as set-up of a semi-continuous biohydrogen production process.

Acknowledgements

Grateful acknowledgements are made to the Natural Sciences and Engineering Research Council of Canada (NSERC) and the Eugenie Ulmer Lamothe Fund (EUL) for their financial support. The analytical assistance of Mr. Ranjan Roy is also gratefully acknowledged.

References

[1] Energy Outlook 2006; Energy Information Administration, Online. http://www.eia.doe.gov/oiaf/aeo/

[2] World Energy Report 2005; RWE, Online www.rwe.com

[3] Shein, K.A., ed., State of the Climate in 2005. *Bulletin of the American Meteorological Society*. **87**: p. S1–S102, 2006.

[4] McHugh, K., *Hydrogen Production Methods*. MPR Associates Inc., 2005.

[5] Duwe, M., *Hydrogen Technology Overview*. Climate Action Network Europe, 2003.

[6] Koku, H., Erlogu, I., Gunduz, U., Yucel, M. & Turker, L., Kinetics of biological hydrogen production by the photosynthetic bacterium Rhodobacter sphaeroides O.U. 001. *International Journal of Hydrogen Energy*, **28(4)**, pp. 381–388, 2003.

[7] Hawkes, F.R., Dinsdale, R., Hawkes, D.L. & Hussy, I., Sustainable fermentative hydrogen production: challenges for process optimisation. *International Journal of Hydrogen Energy*, **27(11–12)**, pp. 1339–1347, 2002.

[8] Hallenbeck, P.C. and J.R. Benemann, Biological hydrogen production; fundamentals and limiting processes. *International Journal of Hydrogen Energy*, **27(11–12)**, pp. 1185–1193, 2002.

[9] Kapdan, I.K. & Kargi, F., Bio-hydrogen production from waste materials. *Enzyme and Microbial Technology*, **38(5)**, pp. 569–582, 2006.

[10] Yetis, M., Gunduz, U., Eroglu, I., Yucel, M. & Turker, L., Photoproduction of hydrogen from sugar refinery wastewater by Rhodobacter sphaeroides O.U. 001. *International Journal of Hydrogen Energy*, **25(11)**, pp. 1035–1041, 2000.

[11] Yu, H., Zhu, Z., Hu, W. & Zhang, H., Hydrogen production from rice winery wastewater in an upflow anaerobic reactor by using mixed anaerobic cultures. *International Journal of Hydrogen Energy*, **27(11–12)**, pp. 1359–1365, 2002.

[12] Kim, M.-S., Baek, J-S., Yun, Y-S., Jun Sim, S., Park, S. & Kim, S-C., Hydrogen production from Chlamydomonas reinhardtii biomass using a two-step conversion process: Anaerobic conversion and photosynthetic fermentation. *International Journal of Hydrogen Energy*, **31(6)**, pp. 812–816, 2006.

[13] Fang, H.H.P., Zhu, H. & Zhang, T., Phototrophic hydrogen production from glucose by pure and co-cultures of Clostridium butyricum and Rhodobacter sphaeroides. *International Journal of Hydrogen Energy*, **31(15)**, pp. 2223–2230, 2006.

[14] Box, G.E.P.H., Hunter, W.G. & Hunter, J.S., *Statistics for Experimenters: An Introduction to Design, Data Analysis and Model Building*, New York: Wiley, 1978.

[15] Owen, W.F., Stuckey, D.C., Healy, J.B., Young, L.Y. & McCarty, P.L., Bioassay for monitoring biochemical methane potential and anaerobic toxicity. *Water Research*, **13(6)**, pp. 485–492, 1979.

[16] Oh, S.E., Van Ginkel, S. & Logan, B.E., The relative effectiveness of pH control and heat treatment for enhancing biohydrogen gas production. *Environmental Science and. Technology*, **37(22)**, pp. 5186–5190, 2003.
[17] Sung, S., Raskin, L., Duangmanee, T., Padmasiri, S. & Simmons, J.J., Hydrogen production by anaerobic microbial communities exposed to repeated heat treatments. *Proceedings of the 2002 U.S. DOE Hydrogen Program Review*, 2002.
[18] Shin, H.-S., Kim, S.-H. & Paik, B.-C., Characteristics of hydrogen production from food waste and waste activated sludge. *Journal of Water and Environment Technology*, **1(2)**, pp. 177–187, 2003.
[19] Khanal, S.K., Chen, W.H., Li, L. & Sung, S., Biological hydrogen production: effects of pH and intermediate products. *International Journal of Hydrogen Energy*, **29(11)**, pp. 1123–1131, 2004.
[20] Chen, W.-H., Chen, S-Y., Khanal, S. & Sung, S., Kinetic study of biological hydrogen production by anaerobic fermentation. *International Journal of Hydrogen Energy*, **31(15)**, pp. 2170–2178, 2006.
[21] Cheong, D.-Y., Hansen, C.L. & Stevens, D.K., Production of bio-hydrogen by mesophilic anaerobic fermentation in an acid-phase sequencing batch reactor. *Biotechnology and Bioengineering*, **96(3)**, pp. 421–432, 2007.

Comparison of three options for biodiesel production from waste vegetable oil

A. A. Refaat & S. T. El Sheltawy
Cairo University, Egypt

Abstract

Biodiesel production deserves continued study and optimization of production procedures due to its environmentally beneficial attributes and its renewable nature. From a waste-management standpoint, production of biodiesel from used cooking oil is environmentally beneficial since it provides a cleaner way of disposing of these products than is typically the case. Biodiesel produced by base-catalyzed transesterification of vegetable oil is usually performed in batch reactors where the required energy is provided by heating accompanied by mechanical mixing. Using this technique, the best yield percentage was obtained using a methanol/oil molar ratio of 6:1, potassium hydroxide as the catalyst (1%), and 65°C temperature for one hour. Alternatively, ultrasonication can provide an effective way to attain the required mixing while providing the necessary activation energy. It was concluded that transesterification by low frequency ultrasound (20 kHz) offered a lot of advantages over the conventional classical procedure. It proved to be efficient (biodiesel yield up to 98–99%), as well as time and energy saving (dramatic reduction of reaction time to 5 min, compared to one hour or more using conventional batch reactor systems, and a remarkable reduction in static separation time to 25 min, compared to 8 hours). The third option studied was the application of microwave irradiation. The application of radio frequency microwave energy offered a fast, easy route to this valuable biofuel with advantages of enhancing the reaction rate and improving the separation process. The methodology allowed for the use of high FFA content feedstock, including used cooking oil, hence reducing the cost of production, which constitutes a major hurdle towards widespread commercialization of biodiesel.

Keywords: biodiesel, waste vegetable oil, transesterification, optimization, ultrasonication, microwave.

1 Introduction

The three basic methods of ester production from oils/fats are base-catalyzed transesterification, acid-catalyzed esterification, and enzymatic catalysis. The most commonly used method of which is the base-catalyzed transesterification technique as it is the most economical process [1]. Transesterification is an equilibrium reaction and the transformation occurs essentially by mixing the reactants. However, the presence of a catalyst accelerates considerably the adjustment of the equilibrium [2]. Methanolysis with 1% wt. potassium hydroxide catalyst resulted in successful conversion giving the best yields and viscosities of the esters [3].

The stoichiometric reaction requires 1 mol of a triglyceride and 3 mol of alcohol. However, an excess of alcohol is used to increase the yields of the alkyl esters and to allow its phase separation from the glycerol formed [4]. The optimum methanol/oil molar ratio was found to be 6:1 [3].

While transesterification is well-established and becoming increasingly important, there remains considerable inefficiencies in existing transesterification processes. In conventional heating of a transesterification process (batch, continuous, and super critical methanol process), heat energy is transferred to the raw material through convection, conduction and radiation from the surfaces of the raw material. Thus, the conventional heating consumes more energy and take long preheat and reaction time, optimally 1 hour [3], to produce over 95 percent conversion yield biodiesel product.

Since this reaction can only occur in the interfacial region between the liquids and also due to the fact that fats and alcohols are not totally miscible, transesterification is a relatively slow process. As a result, a vigorous mixing is required to increase the area of contact between the two immiscible phases and thus to produce an emulsion [5]. Previous research has indicated that ultrasonication provides the mechanical energy for mixing and the required energy for initiating the transesterification reaction [6–10]. Low frequency ultrasonic irradiation is a useful tool for emulsification of immiscible liquids [11]. The collapse of the cavitation bubbles disrupts the phase boundary and causes emulsification by ultrasonic jets that impinge one liquid to another [12].

An alternative energy stimulant, “microwave irradiation”, can be used for the production of the alternative energy source, biodiesel. In the electromagnetic radiation spectrum, the microwave radiation region is located between infrared radiation and radio waves. Microwaves (MW) have wavelengths of 1 mm–1 m, corresponding to frequencies between 0.3–300 GHz. In general, in order to avoid interference, industrial and domestic microwave apparatus are regulated to 12.2 cm, corresponding to a frequency of 2.45 GHz, but other frequency allocations do exist [13]. Microwaves, a non-ionizing radiation incapable of breaking bonds, are a form of energy and not heat and are manifested as heat through their interaction with the medium or materials wherein they can be reflected (metals), transmitted (good insulators that will not heat) or absorbed (decreasing the available microwave energy and rapidly heating the sample) [14]. Microwave heating compares very favourably over conventional methods, where heating can be relatively slow and inefficient because transferring energy into a sample depends upon convection currents and the thermal conductivity of the reaction mixture [15].

2 Materials and methods

The complete methodology of the conventional mixing technique was discussed in detail in a previous study by Refaat *et al.* [3] in which two different biodiesel fuels, obtained from waste cooking oils with different previous uses, were investigated and neat sunflower oil was used as a reference. The experiments have been run in triplicate; each set of operation conditions was conducted three times, once with neat sunflower oil used as a reference and twice with waste vegetable oils (WVOs) obtained from two different sources: one collected from a house (domestic waste vegetable oil – DWVO) and the other from a fast-food popular restaurant (Restaurant waste vegetable oil – RWVO). Methanol was the alcohol of choice: Methanol (Analytical) El-Nasr Pharmaceutical Chemicals Co. (ADWIC) Mwt. 32.04 Assay 99.8%. Both KOH and NaOH were used in that study: NaOH pellets purified SISCO Research Laboratories PVT Ltd – India KOH pellets purified Thann-Fransu. The method applied in that study was the alkali-catalyzed transesterification. The transesterification process was studied at two catalyst loadings (0.5% and 1.0% KOH wt./wt.), two reaction temperatures (25 and 65°C) and three alcohol-to-oil molar ratios (3:1, 6:1 and 9:1). Basically, methanol was the alcohol of choice and KOH was used as the catalyst. The production methodology followed was obtained by researching current methods of small-scale production. Potassium or sodium methoxide solution was prepared freshly by mixing a predetermined amount of methanol (20% by weight of oil) with KOH or NAOH (1.0% by weight of oil) in a container. The reaction was carried out for 1–3 h. under reflux at 60–70°C. Stirring was started with the reaction at the moment of adding potassium or sodium methoxide solution. After the predetermined reaction time, the mixture was carefully transferred to a separating funnel and allowed to stand there overnight. The lower layer (glycerol, methanol and most of the catalysts) was drained out. The upper layer (methyl esters, some methanol and traces of the catalyst) was then cleaned thoroughly by washing with warm (50 °C) de-ionized water. The methyl ester was then dried with anhydrous Na_2SO_4.

The optimum parametric conditions obtained from the conventional technique were applied again using ultrasonication and also microwave irradiation in order to compare the three systems. Ultrasonic mixing was performed with an ultrasonic processor (VC 100, Sonics & Materials Inc., USA). This processor uses electric excitation to generate ultrasound, which is transmitted into the liquid mixture via a titanium probe that causes mixing and provides the energy for the transesterification. The processor operates at 100 watts, 20 kHz frequency. The amplitude and the pulse for the reaction can be adjusted from 0–100%. Process parameters, such as amplitude, pulse, operating time were modulated using the control knobs.

Several examples of microwave irradiated transesterification methods have been reported using adapted domestic ovens to use them as flow systems [16] or batch laboratory ovens [17] but only moderate conversions were obtained. A more recent study used homogeneous catalysis, both in a batch and in a flow system [18]. Leadbeater and Stencel [19] reported the use of microwave heating as a fast, simple way to prepare biodiesel in a batch mode. This was followed by a continuous-flow approach allowing for the reaction to be run under atmospheric conditions and performed at flow rates of up to 7.2 L/min using a 4 L reaction vessel [20]. In this study, a scientific microwave with advanced vessel technology was used. This allowed fast vessel heating with homogeneous microwave distribution throughout the cavity. The oven used is Start S (Milestone), Milestone Inc. (USA). The normal pressure glass reactor is complete with a 500 ml flask and reflux condenser. The oven is supplied with a color touch screen controller that enables creation, storage and use of time-vs.-temperature or time-vs.-power reaction profiles. The output microwave power is adjustable up to 1200 watts, controlled via microprocessor. The temperature was adjusted to 65°C, a methanol/oil molar ratio of 6:1 was employed and potassium hydroxide (1%) was used as a catalyst. The oil was preheated to the desired temperature of 65°C using the microwave unit. The alcohol–catalyst mixture was then fed into the flask through the condenser. The output power was adjusted to 500 watts and the mixture was irradiated under reflux applying different reaction times of 30, 60, 90, 120, 150, 180 and 240 seconds.

GC analysis was undergone for all the samples using Agilent 5975 GC/MS System, Agilent Technologies (USA).

3 Results and discussion

The results obtained by applying the conventional technique were discussed in detail in a previous study by Refaat *et al.* [3]. From the results, it was clear that the produced biodiesel fuel was within the recommended standards of the biodiesel fuel with 96.15% yield. Biodiesel produced from waste vegetable oil was comparable in composition, similar in calorific value to biodiesel produced from refined sunflower oil.

A comparison held between the reaction time, static separation time and yield% on using ultrasonic mixing versus mechanical stirring is shown in table 1.

The reaction time was reduced dramatically from 1 hour to 5 minutes (92%), the static separation time was reduced remarkably from 8 hours to 25 minutes (95%) and the yield% was generally higher.

Application of radio frequency microwave energy enhances the reaction rate for the conversion of triglycerides to biodiesel, and drives the reaction equilibrium toward the production of biodiesel. 100% biodiesel yield was obtained by applying microwave irradiation for two minutes compared to one hour with the conventional technique. The static separation time was 25 minutes compared to 8 hours with the conventional technique. Application of radio frequency microwave energy further improves product recovery in the separation of the biodiesel product from alcohol and glycerin in the reaction mixture. The results achieved by applying microwave irradiation were compared to those obtained by applying the conventional technique (table 2). From these results it is evident that using microwave irradiation reduces the reaction time by about 97% and the separation time by about 94%.

Table 1: Comparison between mechanical and ultrasonic mixing at optimum conditions.

	Yield%		Reaction time (min)		Separation time (min)	
Oil	Mechanical mixing	Ultrasonic mixing	Mechanical mixing	Ultrasonic mixing	Mechanical mixing	Ultrasonic mixing
NVO	96.15	98.94	60	5	480	25
WVOD	95.79	98.26	60	5	480	25
WVOR	94.51	97.68	60	5	480	25

Table 2: Comparison between microwave irradiation and the conventional technique.

System	Reaction Time (min)	Separation time (min)	Yield%
Conventional	60	480	96
Microwave	2	25	100

No substantial differences were obtained with the different origin of the oil. So, using microwave irradiation allows the use of high free fatty acid (FFA) content feed stocks, including animal fats and used cooking oils, in existing transesterification processes by promoting the removal of the fatty acid.

Fig. 1 depicts a comparison between the reaction times using the conventional technique, ultrasonication and microwave irradiation. In fig. 2, the same comparison is held regarding the separation times.

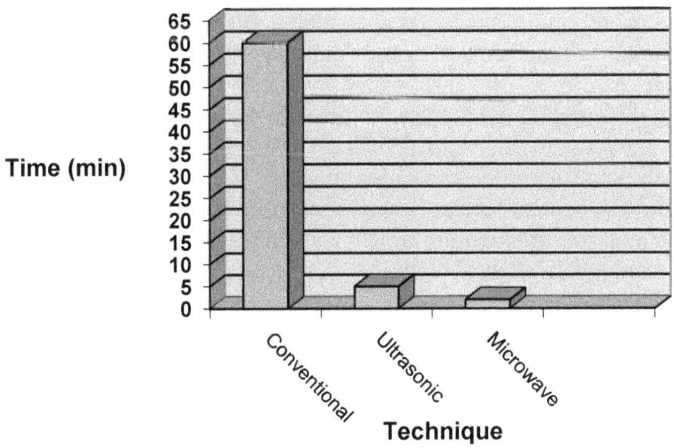

Figure 1: Comparison between the reaction times of the three techniques.

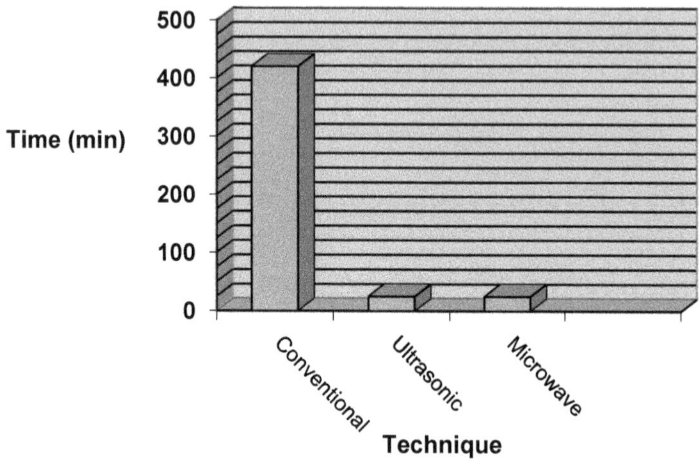

Figure 2: Comparison between the separation times of the three techniques.

4 Conclusions and recommendations

From the present study it can be concluded that low frequency ultrasound is an efficient, time saving and economically functional process, offering a lot of advantages over the conventional classical procedure. The reaction time was reduced dramatically from 1 hr to 5 min (92%), the static separation time was reduced remarkably from 8 hr to 25 minutes (95%) and the yield% was generally higher.

At a frequency of 20 kHz, the optimum reaction time was found to be 5 minutes and the optimum amplitude was found to be 100%. Biodiesel that met the ASTM D 6751 standard was obtained with yields over 98–99% at a separation time of only 25 minutes.

The results showed that the application of radio frequency microwave energy offers a fast, easy route to this valuable biofuel with advantages of enhancing the reaction rate and improving the separation process. The methodology allows for the use of high FFA content feedstock, including used cooking oil; hence it helps to reduce the cost of production, which constitutes a major hurdle towards widespread commercialization of biodiesel.

References

[1] Singh, A.B., He, T.J. & Van Gerpen, J., Process optimization of biodiesel production using different alkaline catalysts. *Applied Engineering in Agriculture*, **22(4)**, pp. 597–600, 2006.

[2] Ma, F. & Hanna, M.A., Biodiesel production: a review. *Bioresource Technology*, **70**, pp. 1–15, 1999.

[3] Refaat, A.A., Attia, N.K., Sibak, H.A., El Sheltawy, S.T. & ElDiwani, G.I., Production optimization and quality assessment of biodiesel from waste vegetable oil. *International Journal of Environmental Science and Technology*, **5(1)**, pp. 75–82, 2008.

[4] Schuchardt, U., Serchelia, R. & Vargas, R.M., Transesterification of vegetable oils: a review. *Journal of Brazilian Chemical Society*, **9(1)**, pp. 199–210, 1998.

[5] Singh, A.K. & Fernando S.D., Catalyzed fast-transesterification of soybean oil using ultrasonication. *American Society of Agricultural Engineers*, ASAE Annual Meeting, Paper number 066220, 2006.

[6] Kelkar, M.A., Gogate, P.R. & Pandit, A.B., Intensification of esterification of acids for synthesis of biodiesel using acoustic and hydrodynamic cavitation. *Ultrasonics Sonochemistry*, **15**, pp. 188–194, 2008.

[7] Stavarache, C., Vinatoru, M. & Maeda, Y., Aspects of ultrasonically assisted transesterification of various vegetable oils with methanol. *Ultrasonics Sonochemistry*, **14(3)**, pp. 380–386, 2007.

[8] Siatis, N.G., Kimbaris, A.C., Pappas, C.S., Tarantilis, P.A. & Polissiou, M.G., Improvement of biodiesel production based on the application of ultrasound: monitoring of the procedure by FTIR spectroscopy. *Journal of the American Oil Chemists Society*, **83(1)**, pp. 53–57, 2006.

[9] Hanh, H.D., Dong, N.T., Starvarache, C., Okitsu, K., Maeda, Y. & Nishimura, R., Methanolysis of triolein by low frequency ultrasonic irradiation. *Energy Conversion and Management*, **49**, pp. 276–280, 2008.

[10] Stavarache, C., Vinatoru, M., Nishimura, R. & Maeda, Y., Fatty acids methyl esters from vegetable oil by means of ultrasonic energy. *Ultrasonics Sonochemistry*, **12**, pp. 367–372, 2005.

[11] Colucci, J.A., Borrero, E.E. & Alape, F., Biodiesel from an alkaline transesterification reaction of soybean oil using ultrasonic mixing. *Journal of the American Oil Chemists Society*, **82(7)**, pp. 525–530, 2005.

[12] Stavarache, C., Vinatoru, M. & Maeda, Y., Ultrasonic versus silent methylation of vegetable oils. *Ultrasonics Sonochemistry*, **13(5)**, pp. 401–407, 2006.

[13] Lidstrom, P., Tierney, J., Wathey, B. & Westman, J., Microwave assisted organic synthesis – a review. *Tetrahedron*, **57**, pp. 9225–9283, 2001.

[14] Varma, R.S., Solvent-free accelerated organic syntheses using microwaves. *Pure and Applied Chemistry*, **73(1)**, pp. 193–198, 2001.

[15] Koopmans, C., Iannelli, M., Kerep, P., Klink, M., Schmitz, S., Sinnwell, S. & Ritter, H., Microwave-assisted polymer chemistry: Heck-reaction, transesterification, Baeyer–Villiger oxidation, oxazoline polymerization, acrylamides, and porous materials. *Tetrahedron*, **62(19)**, pp. 4709–4714, 2006.

[16] Saifuddin, N. & Chua, K.H., Production of ethyl ester (biodiesel) from used frying oil: optimization of transesterification process using microwave irradiation. *Malaysian Journal of Chemistry*, **6(1)**, pp. 77–82, 2004.

[17] Mazzocchia, C., Modica, G., Nannicini, R. & Kaddouri, A., Fatty acid methyl esters synthesis from triglycerides over heterogeneous catalysts in the presence of microwaves. *Comptes Rendus Chimie*, **7(6-7)**, pp. 601–605, 2004.
[18] Hernando, J., Leton, P., Matia, M.P., Novella, J.L. & Alvarez-Builla J., Biodiesel and FAME synthesis assisted by microwaves: Homogeneous batch and flow processes. *Fuel*, **86**, pp. 1641–1644, 2007.
[19] Leadbeater, N.E. & Stencel, L.M., Fast, easy preparation of biodiesel using microwave heating. *Energy and Fuels*, **20**, pp. 2281–2283, 2006.
[20] Barnard, T.M., Leadbeater, N.E., Boucher, M.B., Stencel, L.M. & Wilhite, B.A., Continuous-flow preparation of biodiesel using microwave heating. *Energy and Fuels*, **21**, pp. 1777–1781, 2007.

Characterization of physical and chemical properties of fuel containing animal waste

M. Wzorek
Opole University of Technology, Poland

Abstract

This paper includes the results of research devoted to the application of meat and bone meal as a component of granulated alternative fuel for use in the cement industry. This paper presents the composition of the fuel along with the method of its production. The paper later undertakes an analysis of chemical and physical characteristics of the resulting fuel with particular emphasis on the combustion properties, content of noxious substances and strength prosperities.
Keywords: animal by-products, meat and bone meal, alternative fuels, cement industry.

1 Introduction

Animal by-products generated by the food industry require the application of special treatment techniques due to the hazard they pose to the environment. Because of bacteriological hazard, the intensive, foul smell and difficulties in storing, it is necessary to ensure its fast utilization.

In the utlization plants, the animal by-products undergo the processes of grinding, drying, sterilization and degreasing, which result in the formation of animal meal and fats. In Poland, a total mass of 685,000 tons of animal waste undergoes utilization; however, forecasts say that the volume may reach 770,000 tons within the next 5 to 7 years [1]. The volume of the generated animal by-products in the selected European states is presented in fig. 1.

Depending on the type of the utilized waste, the latter originates from a variety of animal meals: bone meal, meat meal, meat and bone meal, poultry meal and fish meal. The data collected indicate that the output of utilization plants in Poland amounts to around 250 kg of meal and 100 kg of fat out of each ton of animal waste [1].

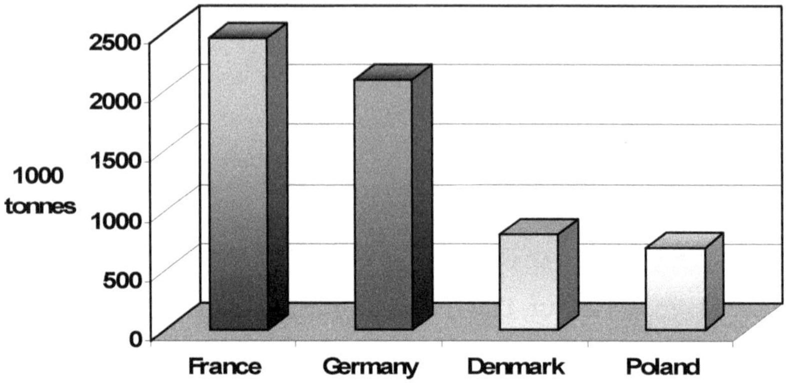

Figure 1: Production of animal by-products in selected European countries [1].

Until now the meat and bone meal (MBM) was mainly applied as a component of feed for farm animals. However, in connection with the threat resulting from the spread of BSE disease and the probability of its effect on humans, the application of meals in feeding farm animals was subjected to strict formal control [2].

Nowadays one disposal option is combustion or the co-combustion of MBM with coal, which ensures that any living organism is totally destroyed and at the same time valorizing its energetic potential [3–6]. The properties which promote the combustion of MBM in industrial processes include the low moisture content, the high calorific value, which is similar to the calorific value of low quality hard coal, and sulfur content lower than that for the case of a majority of brown coal sorts. Properties of animal meal are presented in table 1.

Table 1: Energy parameters of animal waste [7].

Parameter	Unit	Bone meal	Meat meal	Blood meal
LHV	MJ/kg	13.8–14.1	17.3–25.0	21.9–23.4
Moisture	%	3.3–4.9	0.5–7.0	3.9–4.0
Ash	% d.m.	34.5–41.4	10.9–20.5	1.4–1.6
Elementary analysis				
Carbon C	% d.m.	30.5–33.3	40.5–43.8	51.8–52.0
Nitrogen N		3.5–51.8	7.7–10.4	15.6
Hydrogen H		7.3–7.9	4.1–5.3	6.5
Sulphur S		4.4–4.7	0.3–0.5	0.64
Oxygen O		0.10–0.25	13.9–20.0	-

d.m. – in dry mass

However, such factors as the MBM tendency towards displaying powdery characteristics, biological activity and considerable grease content account for the limitations in the application of the meal in the existing installations. Technical problems are associated with the tendency of MBM to form powder and the grease contained in the meal, which results in clogging of the transportation pipes. Biological activity, which is revealed in the conditions of high moisture content, is demonstrated in the processes of decay and hence is associated with health threat to the personnel involved. The variable moisture and fat content in meals and changeable chemical composition result in the necessity to apply specific solutions with regard to combustion conditions and transportation.

This paper includes a presentation of a technique of MBM application to form granulated fuel, which may secure an alternative of an installation in which the input of meal in its powder form may be inadvisable.

2 Materials and methods

As a result of the huge demand for supplementary sources of energy in the form of supplementary fuels, a new fuel designated for application in the process of cement clinker burning resulted from this research.

This fuel applies MBM along with sewage sludge from a municipal waste treatment plant. The application of the other component of the fuel results from the physical properties of the sludge. Since the sewage sludge has a high moisture content, its combination with MBM results in a mixture with moisture content in the range of 40 per cent, which enables its formation into granulate. The properties of the components of the fuel are summarized in table 2.

Table 2: Characteristics of components.

Parameter	Unit	MBM	Sewage sludge
Physical form	-	Grounding ≥4 mm	Pasty consistence
LHV	MJ/kg	17.45	10.73*
HHV	MJ/kg	19.18	11.62
Moisture	%	5.31	71.20
Volatile matter	%	65.21	49.33*
Ash	%	20.46	35.61*

* – in dry mass

The fuel formation was possible in a specially designed drum granulator designed for formation of fuel from waste. The drum is equipped with a special batching system which results in obtaining a granulate diameter in the range from 15 to 35 mm [8]. The method of fuel production consists of the initial mixing of the components at fixed ratios and the subsequent granulation and drying in the drum dryer. The fuel based on bone and meat meal and sewage sludge was named PBM fuel.

The PBM fuel was subsequently subjected to testing aimed at the determination of the combustion and physical properties, including:

a. heating value using Oxygen Bomb Calorimeter KL-Mn,
b. elementary analysis with Elementar Analyzer Vario Macro EL,
c. heavy metal content with ICP MS Perkin Elmer Elan 6100 Mass Spectrometer,
d. chemical composition of ash with Atomic Absorption Spectrometry ASA method,
e. polychlorinated biphenyles (PCBs) with Agilent 6890N gas chromatograph with ECD detector,
f. thermal and thermogravimetric analysis was performed using NETZSCH STA 4009EP differential thermal analyzer within temperature range of 0 ÷ 1400°C in oxygen atmosphere,
g. strength drop in accordance with PN-G-04651: 997 standard,
h. bulk density in accordance with PN-ISO 567:1999 standard.

3 Results and discussion

The application of alternative fuel in the processes of cement linker burning is subjected to numerous technical and technology related restrictions. This is because the fuel supplement with waste must not have a negative effect on the process of cement production in the kiln and, consequently, on the quality of the resulting clinker and the emission levels. In connection with the above, the fuel used in clinker production is required to conform to the required physical and chemical properties.

The properties of the fuel derived from meat and bone meal and sewage sludge was compared with the requirements specified by the cement industry concerning alternative fuels. The most comprehensible effect on the applicability of the fuel based on waste is associated with its calorific value and chemical composition. The energy properties of PBM fuel are summarized in table 3.

Table 3: Energy properties of PBM fuel in comparison to properties of energy industry.

Parameter	Unit	PBM fuel	Requirements stated for cement industry [8, 9]
LHV	MJ/kg	13.55	>13 MJ/kg
HHV	MJ/kg	15. 53	-
Moisture	%	8.67	< 30%
Volatile matter	%	51.92	-
Ash	% d.m.	30.80	< 40%
Elementary composition			
Carbon C		33.47	
Hydrogen H		3.77	
Oxygen O and Nitrogen N	% d.m.	22.67	
Sulphur S		0.62	< 2.5
Chlorine Cl		0.040	< 0.3

d.m. – in dry mass

The analysis of the energy parameters of PBM fuel leads to the conclusion that the properties of the fuel fulfill the minimum requirements of the cement industry. Hence, it is possible to use, as an alternative substitute, PBM fuel at 10 percent of the total volume of fuel use. A greater content of the new fuel may result in a decrease in the kiln effectiveness and increase of the heat used in the process.

PBM fuel contains a considerable amount of ash. However, one has to bear in mind that the ash resulting from the combustion process is absorbed by the clinker and hence it is necessary to take into consideration its chemical composition while deciding about the input of the raw materials in order not to disturb the relations between the constituents forming the material used for preparation of raw mix.

With regard to emissions, the components which are undesirable include sulfur, chlorine, volatile heavy metals (Hg, Cd and Tl) and polychlorinated biphenyls (PCBs), hence, restrictions on their content in alternative fuels have been imposed. The content of the components in the resulting PBM fuel is considerably lower than the admissible levels (tables 3 and 4).

The chlorine content in the meat and bone meal was found equal to around 13% of the level permitted by the cement industry (< 0.3%). For example, chlorine content in the PASi fuel (fuel derived from paper, sawdust and liquid waste, i.e. made from solvents, dyes and varnishes) was equal to 0.24% [9], while in the used car tires (TDF) [10] the content is as much as 0.4%, and in RDF fuel, in the range 0.28 ÷ 0.7% [11]. In comparison to hard coal, the fuel from waste has two times smaller concentration of sulfur (with the values in the range 0.5–3.1% for Polish hard coal, based on the data of ref. [12]).

Table 4: Hazardous material content of in the PBM fuel in comparison to technological restriction imposed on the cement industry.

Parameter	Unit	PBM fuel	Cement industry requirements [8, 9]
Heavy metals	ppm		Total content
Pb		19.43	< 2,500
Zn		446.1	
Sn		0.0301	
Cr		22.08	
Fe		2030	
Cu		43.84	
Ni		4.92	
Cd		0.434	
Tl		0.051	Cd + Tl + Hg < 100
Hg		0.421	< 10
PCB*	ppm	< 0.05	< 50

* – Σ congeners (PCB 28, PCB 52, PCB 101, PCB 138, PCB 180)

The content of volatile metals (Hg, Cd and Tl) is equal to 0.9% in the PBM fuel while the admissible level is 100 ppm. With regard to PCB, the cement industry set the boundary at 50 ppm. Hence, the values of such compounds in PBM fuel are very low (table 4).

At the same time, the detrimental effect on clinker quality may be associated with heavy metals content in ash in the structure of the resulting cement. They are combined into the structure of cement as a result of very durable chemical processes, which may affect clinker structure and hence lead to the deterioration of cement quality.

A number of references used here list information concerning the negative impact of the elementary composition (including Pb, Cb, Zn, Cr, Ni, As, Ag and Ba) on the quality of Portland cement [13–17]. The content of heavy metals in the gained PBM fuel is summarized in table 4.

The highest concentrations of trace metals in the examined fuel was registered with regard to iron, zinc and chromium, which is connected with the high concentrations of these elements in sewage sludge used for the production of the fuel.

Table 5 illustrates the chemical composition of PBM fuel ash in comparison to the composition of clinker and hard coal ash.

Table 5: Chemical composition of PBM fuel in comparison to the components of clinker and hard coal ash.

Compound % of mass	PBM ash	Hard coal ash [18]	Blast furnace slag	Lime stone	Siderite
SiO_2	17.88	42.00	0.70	7.20	32.20
Al_2O_3	4.04	20.00	38.10	2.40	12.20
Fe_2O_3	7.08	17.00	6.90	1.10	27.60
CaO	42.35	5.5	42.50	48.60	2.60
MgO	0.95	2.1	7.10	0.40	2.40
P_2O_5	21.06	-	-	1.5	-
TiO_2	0.36	1.2	-	-	-
SO_3	0.50	5.0	-	0.60	3.20
Compounds not marked					

The main oxides included in PBM fuel ash include: CaO, SiO_2, Fe_2O and Al_2O_3, the list of which is the same as in the raw materials used for clinker production. However, the high content of P_2O_5 is not beneficial since it affects the hydraulic activity of cement. The content of P_2O_5 in clinker must not exceed 0.2–0.3%, hence in the application in PBM fuel it is necessary to pay attention to volume of batched fuel which, besides, must conform to the requirements concerning calorific value and must not display undesired properties.

The experience gained in the cement industry leads to the conclusion that the 20 percent P_2O_5 content in the meal will render it possible to use it for kiln entry in the volume, which corresponds to 5 percent of heat demand.

In order to observe the changes occurring in the fuel under the influence of high temperature, differential thermal analysis (DTA) and thermogravimetric analysis (TG) were undertaken within the range of temperatures from 0 to 1200°C in air atmosphere. The course of TG and DTA analysis is illustrated in fig. 2.

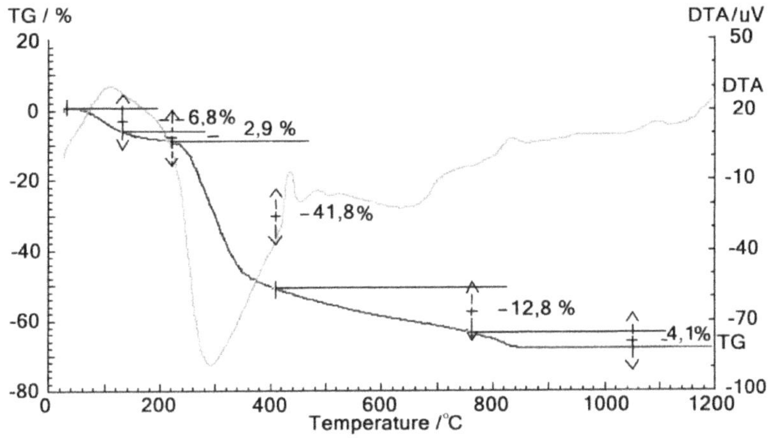

Figure 2: TG and DTA analysis of PBM fuel.

On the basis of analysis of the thermogram, it is possible to conclude that up to a temperature of 150°C, a slight loss of mass follows corresponding to evaporation of the internal moisture. The biggest exothermic peak is noted for a temperature of around 240°C, which is associated with the fuel ignition. The process of combustion occurs up to a temperature of 820°C. The mass loss associated with the burnout of the organic substance is equal to 90%. For a temperature of around 830°C, it is possible to note a slight endothermic effect resulting from the decomposition of calcium carbonate present in the ash.

Apart from the chemical properties of the fuel, it must also fulfill the requirement of ensuring stability, homogeneity of the composition and physical properties.

Some of the other properties, which are particularly relevant to the cement industry, include those which promote its transport, storage and batching. The examinations conducted on the PBM fuel confirm the applicability of the granulated form for application in the kiln. The PBM fuel with 35 mm granulation is shown in fig. 3 and the bulk density is given in table 6.

Stability of the fuel form is extremely important during its transport and storage. This parameter can be assessed by marking its drop strength. This test determines the changes in the structure of fuel when it is dropped from a height of 1.5 m on a concrete foundation under standardized conditions.

Figure 3: PBM fuel.

When fuel is stored, it is often influenced by unfavourable atmospheric conditions, which may negatively affect fuel-hardness. This may be mostly influenced by the change in fuel-moisture as a result of its moisture absorption from the surroundings and precipitation. Therefore, its water resistance is still another specified parameter, which is defined as the degree of change in the drop strength of fuel formed as a result of water absorption.

Table 6: Bulk density of PBM fuel.

Parameter	Unit	Granulation 25 mm	Granulation 35 mm
PBM fuel	kg/m^3	480	325
Biomass pellets [8]		500–780	
Brawn coal briquettes [12]		700–725	

PBM fuel was subsequently subjected to the testing of its resistance to frost. The fuel was subjected to 50 cycles of frosting and defrosting. Each cycle involved freezing the fuel to -18°C, and subsequent defrosting to 18°C. Following that, the resistance of the fuel to drop was tested. The results of this research are presented in fig. 4. The analysis of the results of fuel drop strength confirms that it fully fulfills the necessary strength properties.

On the basis of the conducted research, an observation was made that water has a detrimental effect on the strength of the fuel. It was found that after the fuel absorbs water, its strength tends to drop dramatically. At the same time, the examination of fuel resistance to frost leads to the conclusion that the strength of the fuel after the test fell only by 8%.

The conclusion from the above is that the fuel should be stored in the area protected from rainfall.

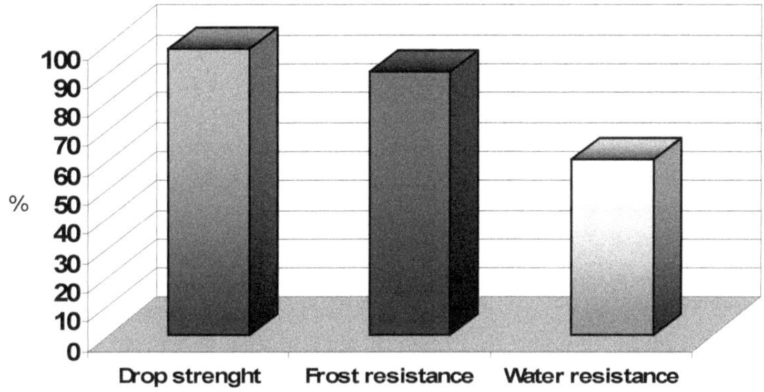

Figure 4: Physical properties of PBM fuel.

The physical properties of briquettes and pellets are to a large degree related to their internal structure. The orientation of grains in the components and binding agent, the occurrence of hollow spaces and cracks and structural faults affect the strength properties of the fuel forming in the process. The assessment of the internal structure of fuels involved the application of computer analysis of the image resulting from the observation of fuel under a microscope (Studar company) and a dedicated software. A sample image with waste and meat and bone meal fuel is presented in fig. 5.

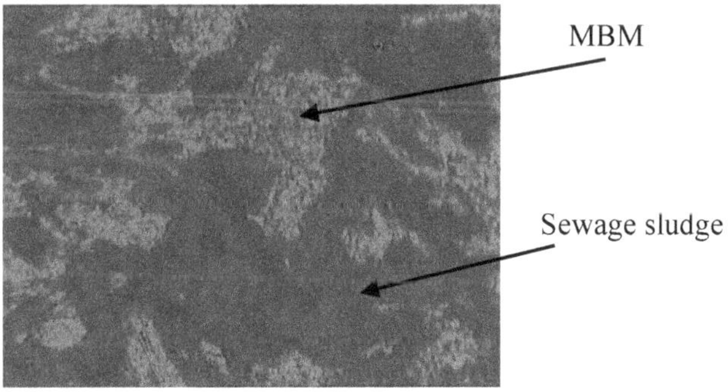

Figure 5: Structure of PBM fuel in 5 times magnification.

By observing the structure of PBM fuel, one can note that the grains in the components the fuel tend to accumulate forming bigger agglomerations.

An important consideration during the production of PBM fuel is associated with good mixing of the components of the fuel since it has an influence on its strength properties (lower drop resistance or too high fuel fragility).

Fuel based on sewage sludge and meat and bone meal has to be manufactured in a plant which is capable to process waste to produce fuel or on a site of a wastewater treatment plant as a final stage of processing sewage sludge to prepare it for energy use.

4 Conclusions

The conducted examinations and analysis leads to the following practical conclusions:

- The application of meat and bone meal as component of granulated fuel may be an alternative for an installation to which the input of animal meal may be undesirable.
- The combined use of meat and bone meal and sewage sludge enables the full application of its energy potential.
- PBM fuel displays properties which fulfill the requirements usually adopted by the cement industry and may be applied as a substitute of coal in the process of clinker cement burning.
- The only problem may be associated with the high content of P_2O_5 in fuel ash, which must be taken into consideration in the determination of maximum P_2O_5 content in the resulting clinker not exceeding the admissible level.
- The properties associated with high water absorption result in the deterioration of the mechanical properties of the fuel. For that reason the fuel should be ensured dry storage facilities.
- The produced fuel can be introduced into the chamber located between the cyclone preheater and the rotary kiln or into the precalciner burner installed in the state-of-the art in modern kiln installations. Those are the places where the kiln can be charged with low-energy fuels showing a high moisture content and likely to contain a larger amount of mineral matter.
- The physical composition and form of the produced PBM fuel makes it applicable for other processes of industrial fuel combustion, e.g., in the process of co-combustion with coal in the existing power plants and heat generating plants.

Acknowledgement

The research was carried out within the project No. 0464/R/2/T02/06/01 funded by the Minister of Science and Higher Education.

References

[1] Schedule of restructuring and modernization of utility industry in Poland Warsaw, 2002 /in Polish/.

[2] Regulation (EC) No 1774/2002 of the European Parliament and of the Council of 3 October 2002 laying down health rules concerning animal by-products not intended for human consumption.

[3] Gulyurtlu, I. & Boavida, D., Abelha, P., Lopes, M.H. & Cabrita, I., Co-combustion of coal and meat and bone meal. *Fuel,* **8(4)**, pp. 2137-2148, 2005.

[4] Skodras, G., Grammelis, P. & Basinas, P., Pyrolysis and combustion behaviour of coal-MBM blends. *Bioresource Technology,* **98**, pp. 1-8, 2007.

[5] McDonnell, K., Desmond, J., Leahy, J.J., Howard-Hildige, R. & Ward, S., Behaviour of meat and bone meal/peat pellets in a bench scale fluidized bed combustor. *Energy,* **26(1)**, pp. 81-90, 2001.

[6] Van Loo, W., *Alternative Fuels: The Valorisation of Waste in the Cement Industry*, CEMBUREAU, Warsaw, 2005.

[7] Wzorek, M., *Experimental Verification of a Model of forming Fuels from Waste based on Example of Municipal Sewage Sludge*, Studies and Monographs, No. 346, Publishing House of Technical University of Opole, Opole, 2013 /in Polish/.

[8] Wzorek, M., Characterization of the properties of alternative fuels containing sewage sludge. *Fuel Processing Technology*, **104**, pp. 80-89, 2012.

[9] Mokrzycki, E. & Uliasz-Bocheńczyk, A., Use of alternative fuels for the cement industry. *Applied Energy*, **74**, pp. 95-100, 2003.

[10] Pipilikaki, P., Katsioti, M., Papageorgiou, D., Fragoulis, D. & Chaniotakis, E., Use of tire derived fuel in clinker burning. *Cement and Concrete Composites*, **27(7-8)**, pp. 843-847, 2005.

[11] Genon, G. & Brizo, E., Perspective and limits for cement kilns as a destination for RDF. *Waste Management*, **28**, pp. 2375-2385, 2008.

[12] Wandrasz, W. & Wandrasz, A.J., *Formed fuels*, Seidel-Przywiecki Publishing House, Warsaw, 2006 /in Polish/.

[13] Murat, M. & Sorrentino, F., Effect of large additions of Cd, Pb, Cr, Zn, to cement raw meal on the composition and the properties of the clinker and the cement. *Cement and Concrete Research*, **28**, pp. 329-333, 1998.

[14] Stephan, D., Mallmann, R., Knöfel, D. & Härdtl, R., High intakes of Cr, Ni, and Zn in clinker Part I. Influence on burning process and formation of phases. *Cement and Concrete Research*, **29**, pp. 1949-1958, 1999.

[15] Trezza, M.A. & Scian, A.N., Waste fuels: their effect on Portland cement clinker. *Cement and Concrete Research*, **35**, pp. 438-444, 2005.

[16] Karagiannis, J., Fitkos, F.J. & Nikolopoulus, P., The use of wastes as alternative fuels in cement production, *Waste Management and the Environment IV*, eds. M. Zamorano, V. Popov, A. Kungolos, C.A. Brebbia, & H. Itoh, WIT Press, Southampton, pp. 105-114, 2008.

[17] Wzorek, M., Baran, T. & Ostrowski, M., The influence of ash absorption from secondary fuels combustion on clinkering process and hydraulic activity of Portland cement clinker. *Cement-Wapno-Beton [Cement-Lime-Concrete],* **4**, pp. 207-215, 2013.

[18] Kikuchi, R., Recycling of municipal solid waste for cement production: pilot-scale test for transforming incineration ash of solid waste into cement clinker. *Resources, conservation and Recycling,* **32**, pp. 137-147, 2001.

The use of wastes as alternative fuels in cement production

J. Karagiannis[1], Ch. Ftikos[2] & P. Nikolopoulos[3]
[1]*Hellenic Cement Research Ctr., Greece*
[2]*National Technical University of Athens, Greece*
[3]*University of Patras, Greece*

Abstract

The high burning temperature, the long retention time inside the kiln, the oxidizing atmosphere, the ash absorption in the clinker and the chemical conditions render the cement kilns ideal installations where wastes can be burned safely. Consequently, some of the wastes may be potential candidates to be used as alternative fuels in cement production that boasts a lot of environmental benefits. These wastes include a variety of materials like used tires, animal meal, refuse-derived fuels (RDF), solvents, used oils and biomass. During the clinker production process, the organic part of the wastes is totally destroyed whereas the inorganic compounds and heavy metals are incorporated in the interstitial phase or substitute calcium or silicon in the silicate phases of the clinker. However, due to their origin these wastes introduce into the cement making process several trace elements that may affect the process itself as well as the properties of the clinker. This work investigates the effects of the use of biomass (cotton stalks) and RDF on the properties of Portland cement clinker. In this respect, ash of these two waste materials was introduced in the raw mix in diverse proportions. The clinkerization characteristics and the microstructure of the lab prepared raw mixes and of the reference mixture were investigated by differential thermal analysis-thermogravimetry, optical microscopy and x-ray diffraction. The obtained results suggest that both waste materials, RDF and cotton stalks can be used as alternatives to fossil fuels, such as coal, for the production of Portland cement clinker.
Keywords: wastes, alternative fuels, Portland cement clinker, trace elements.

1 Introduction

The European cement industry produces annually about 170 million tons of cement. This is produced under rigorous quality specifications imposed by the European and national legislation and relevant internal procedures. Strict regulations apply to cement plants operation which is based upon permission of the pertinent authorities. The cement industry is able to use wastes as alternative fuels in order not only to boost its competitiveness, but also to contribute in the social problem of waste management with an environmental – oriented solution. Today the use of alternative fuels in European countries reaches up to 4 million tons/year, substituting around 2.5 million tons of coal per year [1].

The use of wastes as alternative fuels in cement production boasts a lot of environmental benefits mentioned below:

a. Reduces the use of non-renewable sources like coal as well as the environmental impact of its mining.
b. Contributes towards the reduction of gas emissions responsible for the greenhouse effect by replacing fossil fuels by materials that in another case should have been disposed or burned with the negative impact on the environment.
c. Maximizes energy recovery from wastes. All the energy is used directly within the rotary kiln for clinker production. Also maximizes the recovery of the non-combustible part of the waste (ash) and eliminates any need for its disposal as the remaining inorganic compounds are incorporated in the clinker.

During the clinker production process, the organic part is totally destroyed whereas the inorganic compounds and heavy metals are incorporated in the interstitial phase or substitute calcium or silicon in the silicate phases of the clinker [2–4].

The high burning temperature, the long retention time inside the kiln, the oxidizing atmosphere, the ash absorption in the clinker and the chemical conditions render the cement kilns ideal installations where wastes can be burned safely [5, 6].

Alternative fuels include a variety of materials like used tires, animal meal, refuse-derived fuels (RDF), solvents, used oils and biomass.

The present work, being part of an extensive research program, aims to investigate the properties of clinker burned in laboratory furnace in the presence of RDF and biomass (cotton stalks in this case) used as coal replacements. These materials present a significant calorific value compared to that of coal. Also large amounts of such materials are available domestically to cover the need of cement plants. The mixture design was based upon a 100% replacement of coal by wastes. Therefore, the respective portion of ash of RDF and cotton stalks was introduced in the raw mix. In addition, mixtures containing half and double the amount of the waste input were also examined in comparison. As reference sample (M1C) the same raw mix was used containing conventional coal ash. The clinkering behavior of the initial mixtures was assessed by differential thermal analysis and thermogravimetric analysis (DTA/TG). The resulting clinkers were investigated by X-Ray Diffractometry (XRD) and Optical Microscopy (OM).

2 Experimental procedure

2.1 Materials

The raw mix used in this work was provided by the Cement Company AGET Heracles and corresponds to a typical industrial grey raw mix design. Its chemical analysis performed by means of X-Ray Fluoresence (XRF) is presented in table 1. The mineral investigation of the raw mix carried out by means of XRD technique revealed mainly the presence of calcite ($CaCO_3$) and quartz (a-SiO_2). As secondary phases, dolomite ($CaMg(CO_3)_2$), albite ($NaAlSiO_8$), and illite ($KAl_2(Si_3Al)O_{10}(OH)_2$) were identified. The 90 μm residue was 13.3%.

Table 1: Chemical analysis of the raw mix.

Oxides	(%)	Trace elements	(ppm)	Trace elements	(ppm)
SiO_2	13.29	Tl	15	Ni	40
Al_2O_3	3.55	Pb	22	Hg	0.3
Fe_2O_3	2.28	As	35	Sb	1.4
CaO	42.85	Cd	< 1	Mn	210
MgO	1.37	Co	7	V	42
K_2O	0.38	Cu	39		
Na_2O	0.32	Cr	65		
SO3	0.01	Zn	77		

Table 2: Composition of waste ash.

Oxides	RDF (%)	Cotton stalks (%)	Oxides	RDF (%)	Cotton stalks (%)
SiO_2	24.92	0.64	ZnO	0.188	0.023
Al_2O_3	12.72	2.60	PbO	0.032	0.010
Fe_2O_3	3.32	0.80	MnO	0.076	0.061
CaO	43.68	27.23	CuO	0.056	0.025
MgO	4.40	11.01	NiO	0.050	0.007
SO_3	2.07	7.19	TiO_2	1.940	0.200
K_2O	1.14	29.88	SnO	0.048	traces
Na_2O	3.12	5.32	V_2O_5	traces	traces
Cl	0.80	0.32	CoO	0.010	0.028
Cr_2O_3	0.15	0.04	P_2O_5	5.860	4.760

The RDF was provided by a domestic waste treatment installation. This is a mixture of fabric, plastic and paper which is about 88% combustible. The cotton stalks were provided by local agricultural areas and are about 95% combustible. The chemical analysis of the ash for both waste materials used is presented in table 2. Lower H V and the ash content for these materials is shown in table 3.

Table 3: LHV and ash content of fuels.

Fuel	LHV (J/g)	% Ash content
Coal	30210	9.1
RDF	19672	11.8
Cotton stalks	20260	5.24

2.2 Sample preparation

Based upon industrial data 7.32 kg of coal are needed /100 kg of raw mix. Taking into account the LHV of coal, this means that 2211.4 kJ/kg of raw mix and the respective coal ash residue, namely 0.726 kg, is incorporated into the clinker during burning. Relative to these data and taking into account the LHVs and the ash contents listed in table 3, three basic mixtures were prepared; the reference mixture with coal (M1C) and the mixtures M2R and M2B with a 100% substitution of coal by RDF and cotton stalks respectively. Furthermore, for each one of the wastes used, two more mixtures were prepared that contained half (M1R, M1B) and double (M3R, M3B) of the ash percentage determined for the mixture M2R and M2B. Table 4 illustrates the synthesis and the characteristics of the mixtures prepared.

Table 4: Synthesis and characteristics of the mixtures prepared.

Mixture Name	M1C	M1R	M2R	M3R	M1B	M2B	M3B
Raw mix weight (g)	10	10	10	10	10	10	10
Coal Ash (g)	0.0726						
RDF ash (g)		0.1054	0.2108	0.4216			
Cotton stalks ash (g)					0.0283	0.0566	0.1132
Ash in the raw mix (%)	0.72	1.04	2.06	4.05	0,28	0.56	1.12

The mixture preparation procedure included weighing the portion of the materials (raw mix and ash) and dry mixing. Pellets were prepared from the resulting mixtures by addition of water. The pellets were dried at 100°C and then burned in a programmable furnace, with a rate of 10°C/min, at this temperature for 30 minutes, following air cooling at room temperature.

3 Results and discussion

3.1 Thermal analysis of mixtures

Thermal analysis was performed to simulate the conditions existing in the rotary kiln during clinkering [7–9]. All mixtures were subjected to thermal and thermo gravimetric (DTA/DTG) analyses in air atmosphere. The temperature profile used for heating was identical to that followed for the clinkerization of the pellets in the lab furnace.

The DTA/TG diagrams are presented in fig. 1a for the mixtures M1R, M2R, M3R and (b) for M1B, M2B, M3B in comparison with the reference mixture M1C. All the main reaction peaks observed and the corresponding temperatures are summarized in table 5.

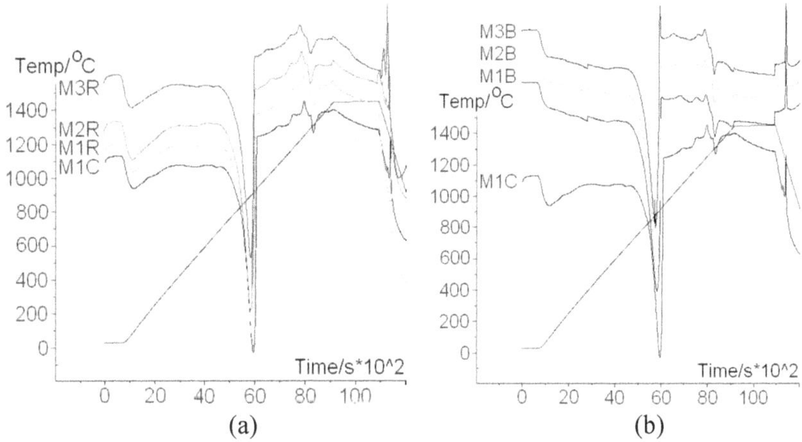

Figure 1: DTA and DTG curves of the reference mixture M1C in comparison to the mixtures containing (a) RDF and (b) cotton stalks ash.

Table 5: Temperatures of the main reaction peaks occurring during the heating of the mixtures.

Mixture	Peak point (°C)		
	Decarbonation	Belite formation	Liquid phase formation
M1C (Reference)	914.7	1249.9	1310.1
M1R	912.7	1243.4	1308.6
M2R	908.0	1238.7	1301.3
M3R	907.2	1228.5	1300.0
M1B	894.6	1248.9	1308.9
M2B	894.5	1243.0	1307.9
M3B	880.8	1237.3	1305.4

The DTA-TG curves clearly show the stage of drying caused by loss of raw mix moisture at about 100°C (broad endothermic peak). At 915°C, a major endothermic peak appears due to the decomposition of calcite, involving loss of carbon dioxide. Right after a very small exothermic peak occurs, connected to the initiation of belite formation. The formation of aluminate and calcium aluminate ferrite is indicated by small endothermic heat effects between 1100 and 1220°C.

The formation of belite is represented by a well formed exothermic peak at 1250°C followed by the endothermic effect of liquid phase formation at 1310°C. The formation of alite occurs above 1350°C and does not show a clear exothermic peak. During the subsequent cooling of the sample, a sharp exothermic peak appears at 1200°C attributed to the crystallization of the molten phase. As revealed by DTA results shown in table 5, being sample, but temperature shifts of the reaction peaks are observed during clinkerization compared to the reference mixture.

Both mixtures containing RDF and cotton stalks ash present lower decarbonation temperatures with increasing ash content. The belite formation and liquid phase formation temperatures are also shifted in the same way.

3.2 X-Ray Diffraction (XRD) study

XRD analyses were carried out by a Rigaku D/MAX-IIVC Geigerflex. All specimens were ground to fine powders. XRD scanning was performed within the range 5°–70° in steps of 0.02° 2θ using Cu*K*a radiation.

The study of the XRD patterns shown in fig. 2, reveals that all the specimens prepared contain all the major clinker phases, namely, alite (C_3S), belite (C_2S), ferrite (C_4AF) and aluminate (C_3A). Table 6 summarises the main peak 2θ position of the two main calcium silicate compounds (C_2S and C_3S). The comparison of the corresponding peaks between the clinkers containing waste ash and the reference one (M1C) does not reveal any significant difference that would suggest any structural modification caused by the trace metal effect contained in the waste ash add.

Table 6: 2θ comparison of the main calcium silicate peaks.

Mixture	**(2θ°)**		
M1C (Reference)	32.200	32.619	34.362
M1R	32.201	32.619	34.374
M2R	32.240	32.600	34.361
M3R	32.200	32.600	34.361
M1B	32.220	32.638	34.363
M2B	32.182	32.600	34.360
M3B	32.201	32.619	34.379

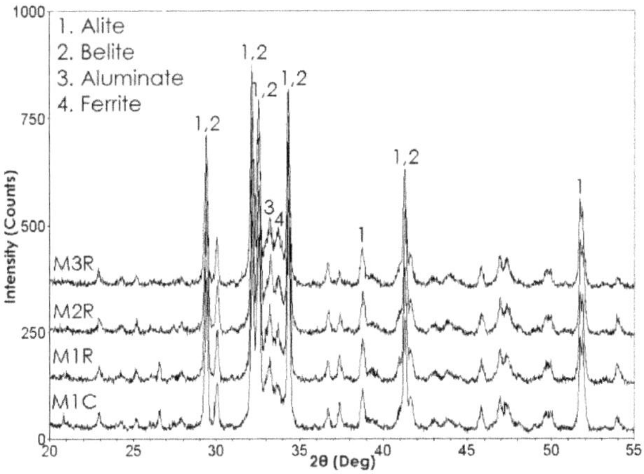

(a)

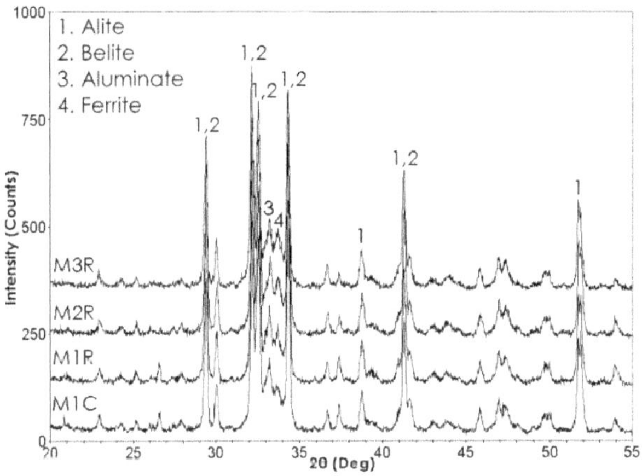

(b)

Figure 2: XRD patterns of the reference mixture M1C in comparison to the mixtures containing (a) RDF (M1R, M2R, M3R), and (b) cotton stalks (M1B, M2B, M3B) ash.

3.3 Optical microscopy (OM) study

Clinkers prepared were mounted in epoxy resin, cut through a cross section and polished. The polished sections were then etched with 40°C de-ionized water, 1% nital and 1% aqueous NH_4CL. Microscopical evaluation was performed in reflected light. Some representative photos of the clinkers examined by means of the optical microscopy are presented in fig. 3.

High porosity was observed in the samples probably due to the preparation method (pelletizing) and clinkering conditions. The numerous belite clusters also observed could be attributed to the same reason [10].

The examination of the reference clinker M1C showed idiomorphic, angular alite crystals with a size between 20–25 μm. The crystals were usually fused. The crystals of belite were in general round with multiple sets of lamellae. Their size was between 25–30 μm. The interstitial phase (aluminate and ferrite) was generally finely crystallized due to the rapid cooling process (figs. 3a and 3b).

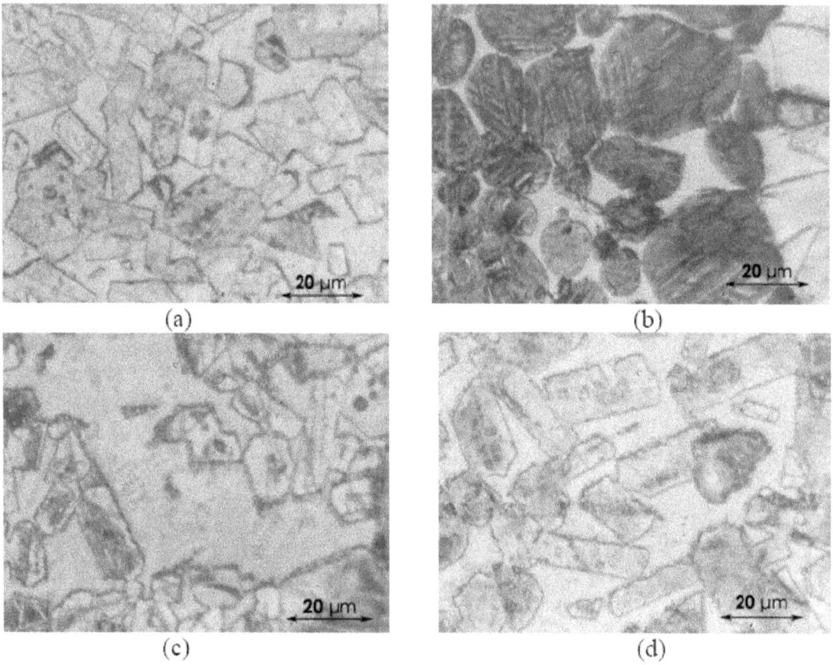

Figure 3: OM photos, (a) clinker M1C showing fused alite crystals of small size, (b) clinker M1C showing round belite crystals with multiple sets of lamellae on their surface, (c) clinker M1R with coarsely crystallized interstitial phase and some alkali aluminate, (d) clinker M1B showing deteriorated alite crystals of small size.

In the case of the clinker with RDF ash addition the alite crystals appeared angular in general, but in some cases belite fringing around their circumference attributed to some decomposition. The crystal size is about 25 μm and partially fused. The belite crystals retained their round shape, size and multilamellae character. The aluminate compounds of the interstitial phase observed in some case affected by alkali incorporation in its lattice thus crystallized in orthorhombic symmetry (fig. 3c).

The clinkers containing ash from cotton stalks have in general similar microstructure to those containing RDF ash. Some of the alite crystals appeared to be deteriorated (fig. 3d). Finally, some alkali aluminates were also observed.

4 Conclusions

Summarising the conclusions derived from the above results, concerned with the effect of the two waste materials used, RDF and cotton stalks, on the clinker formation it is concluded that the clinkers produced contain all the main phases – alite, belite, aluminate and ferrite. The crystal structure of these phases is quite similar to those of the reference clinker. Furthermore, the use of these waste materials has a positive effect on the clinkering by means of promoting the formation of the major clinker phases and shifting their reactions to lower temperatures in comparison to the reference clinker M1C.

These conclusions suggest that the two waste materials used, RDF and cotton stalks, are strong potential candidates to replace fossil fuels like coal for the production of Portland cement clinker. However, further investigations are needed by means of the hydration and strength development of the cement produced with the materials concerned.

Acknowledgements

This project is co-funded by the European Social Fund (75%) and National Resources (25%) – EPEAEK II.

References

[1] Hellenic Cement Industries Association (HCIA), Information available on line at the site *http://www.hcia.gr*

[2] Murat, M. & Sorrentino, F., Effect of large additions of Cd, Pb, Cr, Zn to cement raw meal on the composition and the properties of the clinker and the cement. *Cement & Concrete Research*, **26(3)**, pp. 377–385, 1996.

[3] Kakali, G., Kasselouri, V. & Parissakis, G., Investigation of the effect of Mo, Nb, Wand Zr oxides on the formation of Portland cement clinker. *Cement & Concrete Research*, **20(1)**, pp. 131–138, 1990.

[4] Kakali, G., Tsivilis, S. & Tsialtas, A., Hydration of ordinary Portland cements made from raw mix containing transition element oxides. *Cement & Concrete Research*, **28(3)**, pp. 335–340, 1998.

[5] Trezza, M.A. & Scian, A.N., Waste fuels: their effect on Portland cement clinker. *Cement and Concrete Research*, **35(3)**, pp. 438-444, 2005.

[6] Ract, P.G., Espinosa, D.C.R. & Tenorio, J.A.S., Determination of Cu and Ni incorporation ratios in Portland cement clinker. *Waste Management*, **23**, pp. 281–285, 2003.

[7] Kakali, G., Parissakis, G. & Bouras, D., A Study on the burnability and the phase formation of PC clinker containing Cu oxide. *Cement and Concrete Research*, **26(10)**, pp. 1473–1478, 1996.

[8] Handa, S.K. & Raina, S.J., Cement raw mix characterization by differential thermal analysis. *Thermochimica Acta*, **93**, pp. 609–612, 1985.

[9] Ramachandran, V.S., Paroli, R.M., Beaudoin, J.J. & Delgado, A.H., *Handbook of Thermal Analysis of Construction Materials*, pp. 77–89, 2002.

[10] Hills, L.M. & Tang, F.J., *The Effect of Alternate Raw Materials on Clinker Microstructure & Cement Performance: Two Case Studies*, 1992.
[11] Stephan, D., Knöfel, D., Eber, B. & Härdtl, R., Influence of Cr, Ni, and Zn on the properties of pure clinker phases; Part I: C_3S. *Cement and Concrete Research*, **29**, pp. 545–552, 1999.
[12] Stephan, D., Knöfel, D., Eber, B. & Härdtl, R., Influence of Cr, Ni, and Zn on the properties of pure clinker phases Part II. C_3A and C_4AF. *Cement and Concrete Research*, **29**, pp. 651–657, 1999.
[13] Stephan, D., Mallmann, R., Knöfel, D. & Härdtl, R., Influence on burning process and formation of phase. *Cement and Concrete Research*, **29**, pp. 1949–1957, 1999.

Small decentralised thermal power stations for refuse-derived fuel

C. Bülow
BREWA Umweltservice GmbH, Germany

Abstract

A solution to bringing down energy costs and to making a contribution to climate protection is discussed. Small decentralised thermal power stations for refuse-derived fuel (RDF) with a thermal capacity of 35 MW have some advantages in comparison to large plants. The required amount of RDF (approx. 70,000 t/a) is mostly locally available, which reduces the transport costs considerably. The produced electrical energy (up to 8 MW) and the waste heat from the thermal power station can be used in an industrial plant. An industrial plant and a refuse company founded a joint venture and built a thermal power station for RDF in Bremen. They replaced the old coal fired power station with a RDF-plant. Increasing energy costs are replaced by revenue for RDF. The industrial plant is now using the "cheap" energy from the RDF. The refuse company has a reliable customer and stable prices for its waste for long years: a typical win-win situation. After a 3-year operating time the experiences of the RDF thermal power station are positive without exception. The specified emissions are far below the limits. Furthermore, due to the fact that RDF contains a high proportion of biomass (approx. 45%) by changing the fuel from coal to RDF, a contribution to climate protection can be made. If the same amount of biogene content would be land filled, a large amount of methane, which is one of the more dangerous greenhouse gases, would be produced. The RDF thermal power station is built up modularly. The individual components are standardized so that the plant can be built fast and cost-effectively. The specific investment costs per ton of RDF are comparable to that of large plants (> 100 MW). Due to the handling of RDF in a closed hall, there is no smell outside the plant. The RDF is burnt at over 870°C to eliminate all pollutants completely. The steam is led through a back-pressure turbine and then used in the industrial plant. The modern flue gas treatment plant shows the expected consumption of additives. Since the initiation, the emission limits are constantly considerably below the requested limits of the EU-DIRECTIVE 2000/76/EC.

1 Alternative energy supply through RDF heat power plants

When municipal solid waste is deposited in landfills, methane is generated due to the anaerobic decomposition of the biogene content of waste. Methane is a strong greenhouse gas and about 21 times more powerful at warming the atmosphere than carbon dioxide (CO_2). According to the EU landfill directive (99/31/EC) all member states have to reduce their proportion of the waste for land filling considerably in the following years. They have to sort the waste before land filling and to separate out energy-rich fractions for energetic utilization. Currently most of the waste is still being land filled in the EU.

A lot of the EU-member states initially invest in mechanical biological treatment plants. A utilization of the separated middle- and high caloric fractions often does not exist. Due to the high investment costs, it is often not possible to finance an energetic utilization. But due to the highly increased energy costs, the usage of these fractions as fuel is getting more and more interesting. This especially applies to industrial companies with a high year-round consumption of steam, in particular from the paper-, food- or chemical industry.

This method of cheap and efficient power supply through decentralised small RDF-power stations is increasingly used in Germany. Small decentralised thermal power stations for RDF with a thermal capacity of approx. 35 MW have some advantages in comparison to large plants. The required amount of RDF (approx. 70,000 t/a) is mostly locally available, which reduces the transport costs considerably. The specific investment costs per ton of RDF are comparable to that of large plants (> 100 MW).

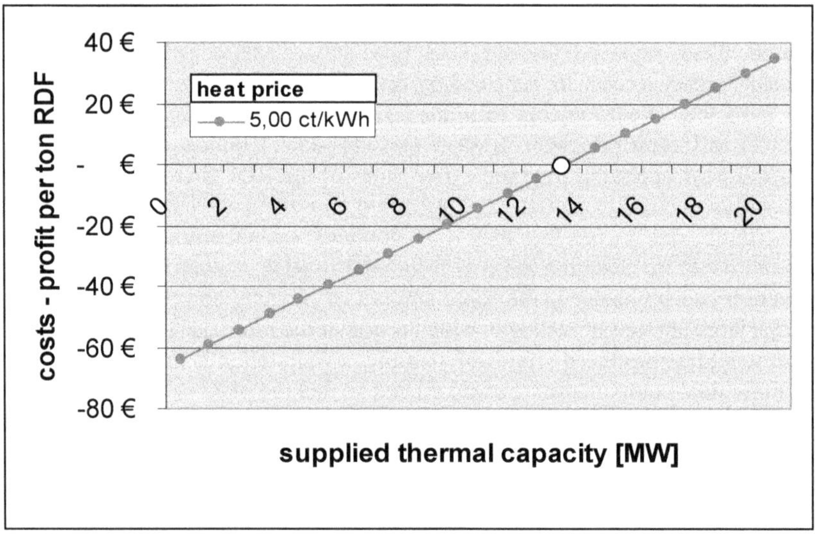

Figure 1: Costs or profit per ton RDF according to the heat supply and price.

2 HKW Blumenthal – a positive example for a RDF power station

An industrial company in Germany, the Bremer Woll-Kämmerei AG (BWK), use the “cheap” energy from the RDF. Until 2004 the BWK had an old coal fired power station to produce electricity and low pressure steam for their production. Increasing repair and fuel costs made it necessary to find a cheaper alternative.

On the other hand, there was a waste management company (The AWG) that operates a mechanical-biological waste treatment plant and a land fill. A German law, based on the EU-Directive, made it necessary to sort the waste before dumping and to separate out energy-rich fractions for thermal utilization.

BWK and AWG founded a joint venture and built a thermal power station for about 60,000 tons RDF per year in Bremen in 2005. The old coal fired power station was replaced with a RDF-plant. The industrial plant is now using the “cheap” energy from the RDF. The refuse company has a reliable customer and stable prices for its waste for long years: a typical win-win situation.

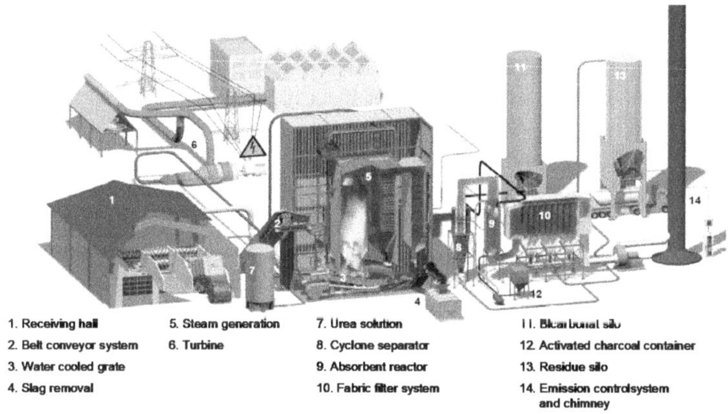

Figure 2: RDF-power plant at the BWK (Germany).

2.1 Technical description

The closed transport containers are emptied in boxes. The walking floor conveyor moves the RDF to the back of the box. The RDF is loosened up and dosed to a belt conveyor. Thus different qualities are mixed to a homogenous fuel. The air of the hall is extracted, burned in the vessel and finally cleaned. The conveyor feeds the RDF to the filling hopper of the firing system.

A hydraulic ram feeds the RDF into the combustion zone of the water-cooled grate system. A multiple zone supply of the combustion air ensures complete and proper combustion.

After complete incineration, the slag is the solid residue of the RDF. It is filled into closed containers and disposed of at an accredited dump site.

The heat of combustion is available for steam generation. Water runs through several pipes (the walls of the vessel also consist of pipes), boils and evaporates. The steam is separated from the remaining water, superheated and is then ready for use.

The steam actuates a turbine for power generation. The steam is led through a back-pressure turbine and then used in the industrial plant.

2.2 Flue gas treatment plant

Urea solution is the initial chemical which supports the flue gas treatment. In a selective non-catalytic process, the nitrogen oxides react to harmless nitrogen (SNCR-process). The cyclone separates coarse dust and its hazardous substances from the flue gas.

For cooling purposes, water is atomised in the absorber reactor. The water evaporates completely. Thus a cooling effect is achieved. The system stays dry. Furthermore, some more chemicals are dosed here for flue gas treatment. Sodium carbonate absorbs acidic components such as sulphur dioxide or hydrogen chloride. Activated charcoal binds possible traces of dioxin, furan, PAH, as well as mercury.

Fabric filter elements separate out the reaction products of the absorber reactor. The reactions of the FGT are finalized in the filtration layer on the fabric filter. The filter system removes the fine dust from the flue gas. The flue gas is totally purified by passing through the filters.

An induced-draft blower exhausts the flue gas from the FGT system and produces a slight under pressure. The flue gas is released to the environment via a chimney of the required height. Automated measuring systems continuously control the clean gas. Violation of limit values sets off an alarm or leads to a plant cut-off.

3 Experience with the RDF-plant

After a 3-year operating time, the experiences of the RDF thermal power station are positive without exception. The RDF is burnt at over 870 °Celsius to completely eliminate all pollutants. The modern flue gas treatment plant shows the

Table 1: Emissions of the RDF-power station 2007.

	Limit value 17.BImSchV* (average per day)	**Average per day 2007**
Nitrogen oxide	200 mg/m^3	168 mg/m^3
Sulfur oxide	50 mg/m^3	11.6 mg/m^3
Dust	10 mg/m^3	0.7 mg/m^3
Carbon oxide	50 mg/m^3	20.3 mg/m^4
Carbon	10 mg/m^3	0.6 mg/m^3
Hydrogen chlorine	10 mg/m^3	7.6 mg/m^3
Heavy metal	0.5 mg/m^3	0.001 mg/m^3
Dioxine	0.1 ng Teq/m^3	0.001 ng Teq/m^3

* German law - dependence on EU-directive 2000/76/EC.

expected consumption of additives. Since the initiation, the emission limits are constantly considerably below the requested limits of the EU-DIRECTIVE 2000/76/EC.

4 Advantages for environment

Due to the fact that RDF contains a high proportion of biomass (approx. 45%) by changing the fuel from coal to RDF, a contribution to climate protection can be made. If the same amount of biogene content would be land filled, a large amount of methane, which is one of the more dangerous greenhouse gases, would be produced. Furthermore fossil fuel is replaced by fuel with high biogene content. CO_2-emission from biomass combustion will account to cero dependence on the rules of the Kyoto protocol.

According to the rules for "clean development projects" and according to the biogene content of the RDF, the new power plant in Bremen can reduce more then 100,000 tons CO_2-emissions (CO_2-equivalents) per year. For equivalent projects in development countries this is an additional possibility to finance "waste to energy" projects.

Sustainable waste processing in a grate furnace and in a fluidized bed incinerator: WtE, recycling and environmental concerns

I. Vermeulen[1], J. Van Caneghem[1], C. Block[1,2], A. Van Brecht[3], G. Wauters[3] & C. Vandecasteele[1]
[1]*Department of Chemical Engineering, University of Leuven, Belgium*
[2]*Leuven Engineering College Groep T, Belgium*
[3]*Indaver NV, Belgium*

The Indaver integrated grate furnace, incinerating municipal solid waste (MSW) along with comparable industrial waste, is described. In the installation, energy is recovered by producing steam which is delivered to other companies, or used to generate electricity. The bottom ashes are wet-washed; ferrous and non-ferrous metals and granulates are recovered. Next to the grate furnace, a fluidized bed combustor (FBC) operated by SLECO is situated. It can co-incinerate various types of industrial wastes (including ASR), RDF, waste water treatment (WWT) sludges, etc. and produces steam to generate electricity. The bottom ashes are recovered as secondary raw material. It is demonstrated that both installations have a good environmental performance and address many aspects of cleaner production. This way, both grate furnace and FBC may play an important role in sustainable waste management. Depending on the fractions of the energy carrier(s), the actual energy recovery varies from 41% for the grate furnace (steam + electricity) to 27% for the FBC (only electricity). The most important airborne emissions and solid residues are monitored in both installation and are discussed in detail. For all components of interest, emissions remain well below Flemish limit values. Moreover, it was shown that both installations act as a POP sink when flue gas emissions are taken into account as a POP output. From the bottom ashes of both incinerators ferrous and non-ferrous metals and granulates are recovered, representing 19.9 and 9.2 wt% of the original waste input of respectively the grate furnace and the FBC. When introducing higher amounts of heavy metals into the FBC, co-incinerating ASR, the bottom ashes still fulfil Flemish requirements for use as secondary raw material.
Keywords: waste-to-energy, material recovery, ASR, MSW, RDF, heavy metals, POPs.

1 Introduction

Sustainable waste management has the objective to deal with waste in an environmentally, economically and socially acceptable way. In this context, incineration of various waste types such as MSW, RDF, WWT sludge, ASR, etc. represents an important treatment option, as it enables recovery of materials and energy, while considerably reducing the amount of waste to be landfilled [1, 2]. The relative importance of different waste treatment and disposal options can vary substantially, depending on the geographical region, as it depends strongly on the specific waste management strategy as well as the availability of space for final landfill disposal [3]. Given that Flanders is very densely populated and due to the stringent European landfill legislation, a lot of effort has been given to reduce the amount of landfilled waste [4].

Indaver is one of the largest waste management companies within the Benelux. Two important waste treatment facilities at the Indaver site, the grate furnace incineration plant and the SLECO fluidized bed combustor, are considered within the present paper. Both incinerators will first be discussed briefly. Next to this, to what extent these installations comply with the relevant aspects of cleaner production will be investigated; such aspects include energy recovery, material recovery and minimal emissions. As waste incinerators are still often considered as a source of POPs, particularly of PCDD/Fs, this aspect will be discussed in more detail. A previously proposed methodology is applied in order to find the aggregated sum of all POPs in the input and in the output of the incinerators. These weighed sums (input vs. output) are subsequently compared in order to investigate whether the considered incinerators must be seen as a POP sink or source. Moreover, the recovery of the bottom ashes is put into focus. From the bottom ashes of both incinerators ferrous metals, non-ferrous metals and granulates are recovered. It is investigated whether the leaching of the bottom ashes still complies with regulatory limitations when higher amounts of heavy metals are introduced into the FBC. It will be demonstrated that both installations combine maximum material and energy recovery with a maximum attention for environment and safety. For each considered aspect, a complete and quantitative picture will be given.

2 Material and methods

2.1 Installations

Both considered installations are located at the Indaver site, situated in an industrial zone close to Antwerp, in the vicinity of one of the most important petrochemical and chemical complexes in the world.

The grate furnace incineration plant is comprised of three grate furnace lines, with an annual capacity of roughly 402×10^3 ton (for the year 2010) [5]. The different components of a grate furnace line are shown in fig. 1. The waste, typically MSW or comparable industrial waste, is delivered in a secluded place (bunker). Using a crane, the waste is introduced onto the moving grate where it is incinerated at a minimum temperature of 850°C. The flue gas is passed through a boiler in order to recover its energy as superheated steam (40 bar, 400°C), which is partly sent to a turbine for electricity production. The flue gas treatment consists of a half-wet spray dryer, injecting lime milk, followed by a baghouse filter and a wet scrubbing installation, using limestone and lime. Activated carbon is injected immediately after the spray dryer to adsorb dioxins, PAHs and volatile heavy metals such as Hg. The resulting solid residues are denominated bottom ash, boiler ash and flue gas cleaning residue. The bottom ash is mechanically treated in order to recover metals and granulates which can be used as secondary raw material. The boiler ash and the flue gas cleaning residue are landfilled after stabilisation/solidification.

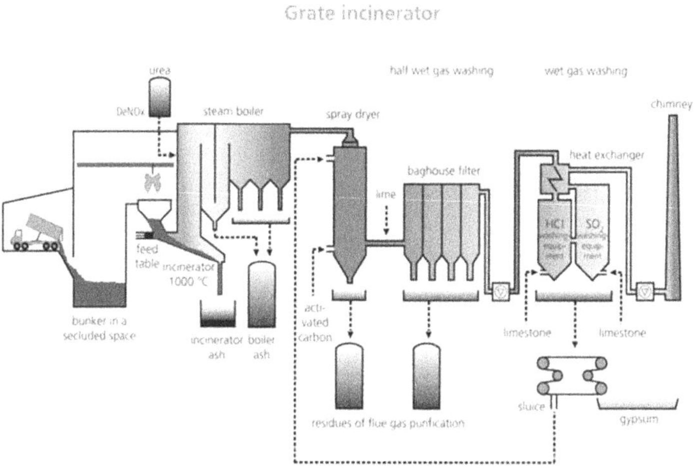

Figure 1: Schematic representation of the grate furnace at Indaver.

The SLECO fluidized bed combustor (FBC) at the Indaver site, generally co-incinerates high calorific RDF with low calorific WWT sludge. This installation has an annual capacity of 500×10^3 ton of mixed waste (for the year 2010), making it the largest and most recent FBC for this purpose in Europe [5]. The FBC is of the ROWITEC internal rotating fluid bed type (fig. 2). The RDF is reduced in size and, after removal of ferrous metals, it is introduced in the fluidized sand bed together with the WWT sludge. During an eight day, full scale experiment in the FBC, 25% of ASR was co-incinerated with the regular waste mix [6–8]. Temperatures in the fluidized sand bed are on average 720°C, while in the freeboard temperatures of about 920°C are reached. Energy from the flue gas is recovered in a vertical waste heat boiler, producing superheated steam (40 bar, 400°C) which is sent to a turbine for electricity production. The flue gas is first dedusted in an electrostatic precipitator (ESP) and flows subsequently through a semi-dry Circoclean reactor, a baghouse filter and a caustic sauda scrubber. A mixture of clay and activated carbon particles (dioxorb) is injected in the semi-dry reactor to ensure efficient PCDD/F removal. At the bottom of the fluidized bed, sand and ashes are removed, after which the sand is sieved of and sent back into the incinerator. The remaining residue, denominated bottom ash, is treated in a recovery plant, producing granulates which can be valorised (useful purpose). Moreover, they fulfil the product criteria according to VLAREA and can therefore be used as e.g. underlayer foundations in road constructions [9]. The other solid residues, fly and boiler ash and flue gas cleaning residue are landfilled after, respectively, moistening and stabilisation/ solidification.

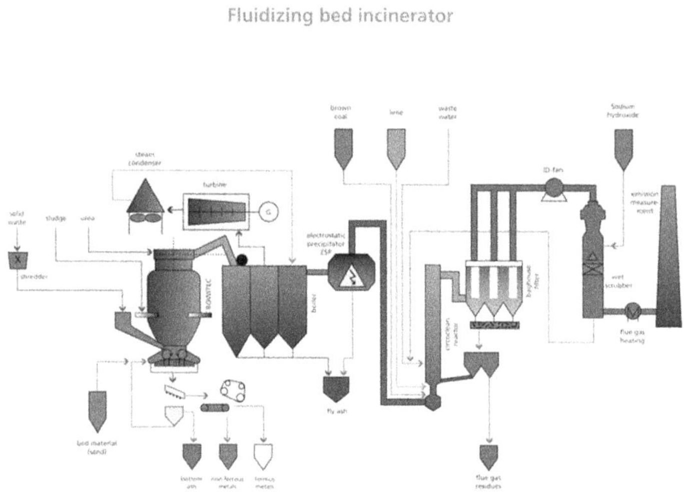

Figure 2: Schematic representation of the FBC at Indaver.

2.2 Sampling and analysis

Flue gases of the different incinerators are monitored, using equipment located on a measuring platform at the stack of the incinerator. Dust is monitored continuously by nefelometry; HCl, SO_2, NO_x and CO by FTIR in order to check their compliance with the regulatory limits. Certain pollutants, such as POPs, cannot be measured continuously; therefore the flue gas is continuously sampled every 2 weeks, using sampling equipment of the AMESA (Adsorption Method for Sampling) type, according to the EN 1948-1 standard. The POPs of interest, e.g. PCDD/Fs, dioxin-like PCBs and PAHs, are subsequently determined by a certified laboratory using GC-MS, according to the EN 1948-2 standard. This semi-continuous way of monitoring is complemented by discontinuous measurements, carried out every 6 months by similar sampling and analysis protocol. Heavy metals in the flue gas are also measured discontinuously. The flue gas is sampled using a quartz filter, analysis is subsequently conducted by using ICP-AES (CMA/2/I/B.1), except for Hg, for which CV-AAS (CMA/2/I/B.3) is used [10]. The heavy metal content of the different solid residues is determined by using EDX-XRF.

3 Aspects of cleaner production and sustainable waste management

3.1 Energy recovery

Energy balance of both grate furnace and fluidized bed combustor at Indaver are summarized in table 1. The total energy input is largely provided by the waste itself; during for instance the start-up phases, fuel oil is used as support. In the grate furnace roughly about 150 ton/h of superheated steam (400°C, 40 bar) is produced, corresponding to a thermal boiler yield of 77.9%; the fluidized bed combustor has a thermal boiler yield of 80.0%. The steam can be used in the installation itself, applied in other installation or used for the production of electricity which is subsequently fed to the electrical grid. The generated energy in the combustion process is thus made available through two energy carriers: steam and electricity. Dependent on the fractions of both carriers, the net energy recovery varies from 41% for the grate furnace (supplying both steam and electricity) to 27% for the FBC (supplying only electricity). Based on the current guidance (29.07.2010) on the Waste Framework Directive 2008/98/EC, both the grate furnace (0.82) and the FBC (0.70) are classified as recovery (R1) [11].

Table 1: Energy balance of the grate furnace and the FBC at Indaver (data 2010) [5].

Grate furnace	**GJ/year**	**MWh/year**
Energy content of input	4,440 10^3	
Energy content of steam produced	3,459 10^3	
Electricity produced	529 10^3	147 10^3
Steam supplied for application as heat	1,309 10^3	
FBC	**GJ/year**	**MWh/year**
Energy content of input	4,557 10^3	
Energy content of steam produced	3,645 10^3	
Electricity produced	1,213 10^3	337 10^3
Steam supplied for application as heat	/	

3.2 Emissions to air

Table 2 gives an overview of the average yearly emission concentrations for all measured elements in both incinerators; for the full scale experiment (FBC, 50%sludge-25%RDF-25%ASR) the average emissions concentrations during the eight day experiment are given. It is clear that for both installations all emissions remain well below regulatory limit values, also when co-incinerating ASR in the FBC. The volatile heavy metals, Hg, Cd and Tl, that are monitored separately, remain up to several orders of magnitude underneath regulatory limit values.

Table 2: Average (yearly) emissions concentration and limit values, expressed in mg/Nm³ (data 2010) [5].

Heavy metal	**Grate furnace**	**FBC 30% sludge-70% RDF**	**FBC 50% sludge-25% RDF-25%ASR**	**Limit value**
Dust	0.84	0.6	0.2	10
CO	14.31	14.0	4.1	50
TOC	1.82	0.9	0.8	10
HCl	0.93	0.1	0.4	10
SO_2	1.55	1.6	0.0	50
NO_x	145	92	109.8	200*
Hg	< 0.0005	0.013	0.0098	0.05
Cd + Tl	0.00011	< 0.016	0.00034	0.05
∑(Sb, As, Pb, Cr, Co, Cu, Mn, Ni, V, Sn)	0.09	0.05	0.021	0.5

*For the FBC the limit value for NO_x is 125mg/Nm³

The average annual emission concentration of PCDD/F for 2009, calculated on the basis of semi-continuous measurements, is 0.013 ngTEQ/Nm³ for the grate furnace and 0.008 ngTEQ/Nm³ for the FBC. This corresponds to a total annual emission of respectively 0.026 gTEQ and 0.019 gTEQ for the grate furnace and the FBC. This total annual emission can be considered negligible compared to the overall estimated emissions for Flanders of 42.3 gTEQ in 2010 [12]. The major emission source of PCDD/F in Flanders is found to be back yard burning, accounting for up to 55% of the total emissions.

Next to this, the total amount of POPs, not just the different groups of POPs separately, in input and output of the incinerators is compared in order to investigate whether the considered installations are a POP sink or source. The authors of the present paper have previously proposed a methodology to determine the aggregated sum of all POPs by weighing the different POPs according to their minimal risk doses (MRDs) or cancer potency factors [6, 7, 13]. For the input of the grate furnace, PCDD/Fs, PBDD/Fs and polybrominated diphenylethers (PBDEs) are the main contributors. The concentration of POPs in RDF, WWT sludge and MSW is comparable, even though their specific POP-fingerprint (distribution of the different POP congeners) can differ considerably [6, 7]. The PCDD/F and PCB concentrations in MSW, RDF and WWT sludge, when expressed in $pgTEQ/g_{dw}$, are roughly 100 times lower as the concentrations found in ASR; the PAHs concentrations are 10 times lower [6, 7]. Table 3 summarizes the ratios of the weighed POP-input over POP-output, for the different incinerators, when weighed with 1/MRDs and cancer potency factor respectively. When for the POP-output, only flue gas emissions are taken into account, the input over output ratios are in all cases significantly larger than 1 for both weighing methods, indicating that the incinerators act as a POP sink. When all outputs are considered, the weighed POP-input is comparable to the weighed POP-output for the grate furnace and the FBC (30%sludge-70%RDF). When however higher amounts of POPs are introduced into the FBC, co-incinerating ASR, the FBC acts in all cases as a weighed POP sink. This can be explained by the fact that the POPs in the waste input are almost completely destroyed during incineration and the formation of new POPs during the cooling of the flue gas seems to a large extent independent of the POP concentrations in the incinerated waste [7].

Table 3: Weighed POP input over output ratios.

	Weighing with (1/MRD)	Weighing with potency factor
Grate furnace		
Total input/output(flue gas)	90–580	6–55
Total input/total output	0.3–3.2	0.1–2.0
FBC 30% sludge-70%RDF		
Total input/output(flue gas)	120–740	40–420
Total input/total output	0.31–0.97	3.2–14
FBC 50% sludge-25%RDF-25%ASR		
Total input/output(flue gas)	9700–13,000	1800–21,500
Total input/total output	5.6–19	9.1–127

3.3 Material recovery

Table 4 summarizes the weight percentages of the solid residues of both grate furnace and FBC at Indaver. The bottom ashes of the grate furnace are wet-washed prior to their recovery. During this treatment ferrous and non-ferrous metal fractions are recovered and the resulting bottom ash is split up into several granulate fractions based on their size. These granulate fractions as well as the recovered metal fractions can be valorised (useful purpose). Moreover, they fulfil the product criteria according to VLAREA and can therefore be used as e.g. underlayer foundations in road constructions [9]. Only the sludge fraction, representing 12 wt% of the bottom ashes of the grate furnace, has to be landfilled.

Table 4: Mass balance of the grate furnace and the FBC at Indaver (data 2010) [5].

Grate furnace	**wt% of input**
Recovered metal fraction	2.9
Bottom ash	22.6
Boiler ash	1.9
FGC residue (including gypsum)	2.8
Total to be landfilled	**7.5**
Fluidized bed combustor	**wt% of input**
Recovered metal fraction (scrap)	3.2
Bottom ash	6.0
Fly and Boiler ash	10.6
FGC residue	2.0
Total to be landfilled	**12.6**

The bottom ashes of the FBC are completely recovered as secondary raw material. These ashes are treated in a bottom ash recovery plant, where ferrous and non-ferrous metals are recovered and granulates are produced which can be used in road construction. The other solid residues of both installations: (fly and) boiler ash and flue gas cleaning residue are landfilled, after stabilisation/solidification. All in all only 7.5 wt% of the original waste input of the grate furnace and 12.6% of the original waste input of the FBC needs to be landfilled.

The bottom ashes, or at least a considerable fraction of the bottom ashes, of both installations are thus recovered as secondary raw material. Table 5 gives the leaching results of the recovered bottom ashes of both installations. For the FBC a distinction is made between the bottom ashes when incinerating the normal waste input (30%sludge-70%RDF) and when co-incinerating 25% of ASR. It can be seen that all leaching results are well below Flemish requirements (VLAREA limit value), even when ASR is co-incinerated in the FBC (Table 5). In the latter case however, a slight increase in the leaching results of several heavy metals, including Cu, Ni, Pb and Zn, is found. This increase can be attributed to ASR, containing higher concentrations of certain heavy metals than the other input streams (WWT sludge and RDF) [8].

Table 5: Metal leaching data from bottom ashes of grate furnace and FBC at Indaver (data 2007) [5].

Heavy metal	Grate furnace	FBC 30%sludge-70%RDF	FBC 50%sludge-25%RDF-25%ASR	VLAREA Limit value
As	< 0.05	< 0.02	0.08	0.8
Cd	< 0.01	< 0.002	< 0.002	0.03
Cr	< 0.05	0.49	0.01	0.5
Cu	0.29	0.02	0.17	0.5
Pb	< 0.10	< 0.01	0.53	1.3
Ni	< 0.05	< 0.01	0.06	0.75
Zn	< 0.10	0.01	0.03	2.8

4 Conclusions

It was demonstrated that both considered installations have a good environmental performance and address many aspects of cleaner production: energy recovery, material recovery and minimal emissions. This way, both grate furnace and FBC may play an important role in sustainable waste management. Both installations offer good energy recovery; depending on the energy carrier(s) this can vary from 41% for the grate furnace (supplying both steam and electricity) to 27% for the FBC (supplying only electricity). For all components of interest, emissions to air remain well below regulatory limit values. Moreover, it was shown that both installations act as a sink of POPs when only flue gas emissions are taken into account as POP output. The total amount of weighed POPs in the outputs, when considering all fractions, was found to be comparable to the total amount of weighed POPs in the input for the grate furnace and the regular waste input of the FBC. When introducing higher amounts of POPs in the input of the FBC, co-incinerating ASR, ratios of POP-input over-output increase so that the FBC can in all cases be considered as a POP sink.

From the bottom ashes of both incinerators ferrous metals, non-ferrous metals and granulates (subdivided based on their size) are recovered, representing 19.9 and 9.2 wt% of the original waste input of respectively the grate furnace and the FBC. Even when higher amounts of heavy metals are introduced into the FBC, co-incinerating ASR, leaching results of the bottom ashes still comply with regulatory limit values. All in all, only 7.5 wt% of the original waste input of the grate furnace and 12.6 wt% of the original waste input of the FBC needs to be landfilled.

References

[1] Vandecasteele, C., Wauters, G., Arickx, S., Jaspers, M. & Van Gerven, T., Integrated municipal solid waste treatment using a grate furnace incinerator: The Indaver case. *Waste Management*, **27**, pp. 1366-1375, 2007.

[2] Vermeulen, I., Van Caneghem, J., Block, C., Baeyens, J. & Vandecasteele, C., Automotive shredder residue (ASR): reviewing its production from end-of-life vehicles (ELVs) and its recycling, energy or chemicals' valorization. *Journal of Hazardous Materials*, **190(1-3)**, pp. 8-27, 2011.

[3] Sabbas, T., Polettini, A., Pomi, R., Astrup, T., Hjelmar, O., Mostbauer, P., Cappai, G., Magel, G., Salhofer, S., Speiser, C., Heuss-Assbichler, S., Klein, R. & Lechner, P., Management of municipal solid waste incineration residues. *Waste Management*, **1**, pp. 61-88, 2003.

[4] Background document on waste management, www.milieurapport.be

[5] Van Brecht, A. Personal communication, May 2011, Innovation Engineer, Indaver, Mechelen, Belgium.

[6] Van Caneghem, J., Block, C., Vermeulen, I., Van Brecht, A., Van Royen, P., Jaspers, M., Wauters, G. & Vandecasteele, C., Mass balance for POPs in a real scale fluidised bed combustor co-incinerating automotive shredder residue. *Journal of Hazardous Materials*, **181**, pp. 827-835, 2010.

[7] Van Caneghem, J., Vermeulen, I., Block, C., Van Brecht, A., Van Royen, P., Jaspers, M., Wauters, G. & Vandecasteele, C., Destruction and formation of PCDD/Fs in a fluidised bed combustor co-incinerating automotive shredder residue with refuse derived fuel and waste water treatment sludge. *Journal of Hazardous Materials*, **207-208**, pp. 152-158, 2012.

[8] Vermeulen, I., Block, C., Dewulf, W. & Vandecasteele, C., Energetic valorisation of automotive shredder residue by co-incineration in a fluidized bed combustor. *Proc. of the 2nd Int. Conf. on Hazardous and Industrial Waste Management*, Section A4, pp. 91-93, 2010.

[9] VLAREA, http://www.emis.vito.be/vlarea

[10] CMA, compendium for sampling and analysis in the frame of the waste and soil sanitation regulation, http://www.emis.vito.be/referentielabo-ovam

[11] European Parliament and the European Council, Directive 2008/98/EC of the European Parliament and of the Council of 19 November 2008 on waste and repealing certain Directives Text with EEA relevance. *Official Journal of the European Communities L*, **312**, pp. 3-30, 2008.
[12] Core Set of Environmental Data MIRA-T, www.milieurapport.be
[13] Van Caneghem, J., Block, C., Van Brecht, A., Wauters, G. & Vandecasteele, C., Mass balance for POPs in hazardous and municipal waste incinerators. *Chemosphere*, **78**, pp. 701-708, 2010.

Numerical optimisation of operating conditions of waste to energy

M. Kapitler, F. Kokalj & N. Samec
Faculty of Mechanical Engineering, University of Maribor, Slovenia

Abstract

The combustion process, using municipal solid waste (MSW) as a fuel in waste to energy plant, calls for a detailed understanding of these phenomena. On the one hand, this process depends on many input parameters, like MSW proximate and ultimate analysis, the season of the year, primary and secondary inlet air velocity and, on the other hand, on the output parameters such as temperature or mass flow rate of combustion products on the exhaust. The variability and mutual dependence of these parameters can be difficult to manage in practice. Moreover, another problem is how these parameters can be tuned to achieve the optimal combustion with minimal pollutants emission during the plant design phase already. To meet these goals, a waste to energy plant with bed combustion is investigated by using a computational fluid dynamics approach with ANSYS CFX 12.0 code in WORKBENCH 2 environment. In this paper, the adequate variable input boundary conditions which are based on the real measurement and practical calculations of known MSW composition from other authors are used and the whole computational work is updated with real plant geometry and the appropriate turbulence, combustion and heat transfer models. Furthermore, the operating parameters were optimized on output parameters through trade-off study. Different operating conditions were varied and the fluid flow direction, residence time, temperature field, velocity field, nitric oxide formation and combustion products through plant combustion chamber and preheat intersection in 3D were predicted and visualized. The optimization in real time has shown the amount of each input parameters to meet the optimal operating conditions. Finally, the response charts between the input and output parameters are presented to monitor the dependence among these parameters. Further simulations have to be done to include the geometry dimensions as input parameters.
Keywords: municipal solid waste modelling, bed combustion, computational fluid dynamics, operating conditions, numerical optimisation, trade-off study, parameter sensitivity, response charts.

1 Introduction

Nowadays, bed combustion on grate is the most common way to burn municipal solid waste (MSW) in waste to energy plants (WTEP). The combustion in these plants are very specific due to the characteristics of fuel (MSW) which depends on proximate and ultimate analysis, season of the year, primary and secondary inlet air velocity and many other parameters which change all the time in some frames. The goals are how the optimal combustion can be tuned with minimal emission of the pollutants in the environment already in the plant project phase. To meet these goals, WTEP with bed combustion is investigated by using computational fluid dynamics (CFD) approach [1–11] with ANSYS CFX 12.0 code in the WORKBENCH 2 environment. By using this tool the input and output parameters can be followed and the mutual interaction visualized. The mathematical models were developed like FLIC [5, 7–10] and test project TAMARA [3] to find out the combustion products distribution in the gaseous phase along the travel bed. The combustion of MSW in the gaseous phase combustion has many phases like moisture evaporation, waste devolatilisation, combustion of volatiles, mixing and fixed carbon combustion during heterogeneous chemical reactions [10].

2 Numerical optimization of operating conditions

2.1 Models description

The combustion in a WTEP is described quite accurately by a system of differential equations of the reacting flow so called Reynolds Averaged Navier-Stokes equations (RANS), presented in the following form:

$$\frac{\partial \bar{\rho}}{\partial t} + \frac{\partial}{\partial x_j}\left(\bar{\rho}\bar{\upsilon}_j\right) = 0 , \tag{1}$$

$$\frac{\partial}{\partial t}\left(\bar{\rho}\bar{\upsilon}_j\right) + \frac{\partial}{\partial x_j}\left(\bar{\rho}\bar{\upsilon}_j\bar{\upsilon}_i\right) = -\frac{\partial p}{\partial x_i} + \overline{f_{\upsilon i}} - \frac{\partial}{\partial x_j}\left(\bar{\tau}_{ij} + \overline{\rho\upsilon_j'\upsilon_i'}\right) , \tag{2}$$

$$\frac{\partial}{\partial t}\left(\bar{\rho}\bar{h}\right) + \frac{\partial}{\partial x_j}\left(\bar{\rho}\bar{\upsilon}_j\bar{h}\right) - \frac{\partial p}{\partial t} + \frac{\partial}{\partial x_j}\left(\bar{q}_j + \overline{\rho\upsilon_j'h'}\right) = \overline{I_T} . \tag{3}$$

Reynolds' stresses ($\overline{\rho\upsilon'_j\upsilon'_i}$) are modelled by the introduction of turbulent viscosity η_t:

$$\overline{\rho\upsilon_j'\upsilon'_i} = \frac{2}{3}\delta_{ij}\left(\rho k + \eta_t\frac{\partial\upsilon_k}{\partial x_k}\right) - \eta_t\left(\frac{\partial\upsilon_i}{\partial x_j} + \frac{\partial\upsilon_j}{\partial x_i}\right) . \tag{4}$$

The turbulent viscosity can be determined using various turbulent models to close-down the system of Reynolds' equations. The two-equation k - ε turbulent model is used for the purpose of the presented reacting flow modelling. The application of k - ε turbulent model in the modelling of reacting flows has already been proven by many authors as a very successful one. The turbulent viscosity is computed using:

$$\eta_t = \rho C_\eta \frac{k^2}{\varepsilon}, \tag{5}$$

where k is turbulent kinetic energy – $k = 0.5\left(\overline{\upsilon_i'\upsilon_i'}\right)$ and ε its dissipation (irreversible transformation of kinetic energy into internal energy).

Local values of k and ε are computed using the following transport equations:

$$\frac{\partial}{\partial t}(\rho k) + \frac{\partial}{\partial x_j}\left(\overline{\upsilon}_j k\right) - \frac{\partial}{\partial x_j}\left[\left(\eta + \frac{\eta_t}{\sigma_k}\right)\frac{\partial k}{\partial x_j}\right] = I_k, \tag{6}$$

$$\frac{\partial}{\partial t}(\rho \varepsilon) + \frac{\partial}{\partial x_j}\left(\overline{\upsilon}_j \varepsilon\right) - \frac{\partial}{\partial x_j}\left[\left(\eta + \frac{\eta_t}{\sigma_\varepsilon}\right)\frac{\partial \varepsilon}{\partial x_j}\right] = I_\varepsilon, \tag{7}$$

the source terms are modelled as:

$$I_k = \eta_t\left(\frac{\partial \overline{\upsilon}_i}{\partial x_j} + \frac{\partial \overline{\upsilon}_j}{\partial x_i}\right)\frac{\partial \overline{\upsilon}_i}{\partial x_j} - \rho\varepsilon, \tag{8}$$

$$I_\varepsilon = C_1 \frac{\varepsilon}{k}\left[\eta_t\left(\frac{\partial \overline{\upsilon}_i}{\partial x_j} + \frac{\partial \overline{\upsilon}_j}{\partial x_i}\right)\frac{\partial \overline{\upsilon}_i}{\partial x_j}\right] - C_2\rho\frac{\varepsilon^2}{k}. \tag{9}$$

Reynolds' enthalpy flux $\overline{\rho\upsilon_j'h'}$ in eqn (3) is also defined with turbulent viscosity:

$$\overline{\rho\upsilon_j'h'} = -\frac{\eta_t}{\mathrm{Pr}_t}c_p\frac{\partial T}{\partial x_j}, \tag{10}$$

where Pr_t is the turbulent Prandtl number. C_η, C_1, C_2, σ_k and σ_ε are constants, and their values used in the presented work are: $C_\eta = 0.09$; $C_1 = 1.44$; $C_2 = 1.92$; $\sigma_k = 1$ and $\sigma_\varepsilon = 1.3$.

Advection – diffusive equation of mass species (ξ_k) of the component k has due to Reynolds' averaging, an additional term called turbulent mass species flux:

$$\overline{\rho\xi_k'\upsilon_j'} = \frac{\eta_t}{\mathrm{Sc}_t}\frac{\partial \xi_k}{\partial x_j}, \tag{11}$$

and can be modelled with turbulent viscosity using the k-ε model. The complete advection – diffusive mass species equation is:

$$\frac{\partial}{\partial t}\left(\overline{\rho \xi_k}\right)+\frac{\partial}{\partial x_j}\left(\overline{\rho \upsilon_j \xi_k}\right)-\frac{\partial}{\partial x_j}\left[\left(\rho D_k+\frac{\eta_t}{\mathrm{Sc}_t}\right)\frac{\partial \xi_k}{\partial x_j}\right]=\overline{I_{\xi_k}}, \tag{12}$$

where Sc_t is the turbulent Schmidt number and D_k molecular diffusion coefficient of component k. With the new term:

$$\Gamma_{k,eff}=\rho D_k+\frac{\eta_t}{\mathrm{Sc}_t}=\Gamma_k+\frac{\eta_t}{\mathrm{Sc}_t}, \tag{13}$$

eqn (12) can be rewritten as:

$$\frac{\partial}{\partial t}\left(\overline{\rho \xi_k}\right)+\frac{\partial}{\partial x_j}\left(\overline{\rho \upsilon_j \xi_k}\right)-\frac{\partial}{\partial x_j}\left(\Gamma_{k,eff}\frac{\partial \xi_k}{\partial x_j}\right)=\overline{I_{\xi_k}} \tag{14}$$

Source terms of energy and mass species transport equations are computed by the following two equations where ω_k is computed from the turbulent combustion model:

$$\overline{I_T}=-\sum_{k=1}^{N}\Delta H^o_{f,k}\overline{\omega}_k, \tag{15}$$

$$\overline{I_{\xi_k}}=M_k\overline{\omega}_k, \tag{16}$$

where $\Delta H^\circ_{f,k}$ is the standard heat formation and M_k the molecular mass of component k. In eqns (15) and (16), ω_k stands for the formation/consumption rate of component k and is defined by the following expression:

$$\overline{\omega}_k=\frac{d[X_k]}{dt}=\left(\nu_k''-\nu_k'\right)\overline{R}_k, \tag{17}$$

which is written for the general form of the chemical reaction:

$$\sum_{k=1}^{N}\nu_k' X_k \underset{k_b}{\overset{k_f}{\rightleftarrows}} \sum_{k=1}^{N}\nu_k'' X_k, \tag{18}$$

where ν'_k and ν''_k represent the stoichiometric coefficients of component k for reactants and products, respectively. Chemical reaction rate R_k in eqn (17) is calculated from appropriate combustion models. It has to be pointed out that nowadays many turbulent combustion models are in practical use. Their application depends on the type of combustion (diffusion, kinetic, mixed), fuel type (solid, liquid, gaseous) and combustion device (furnace, boiler, engine). Most of models include various empirical constants which need to be individually determined case by case. In this case the Eddy Dissipation Combustion Model has been applied [12, 13].

2.2 Geometry, meshing and boundary conditions

With the CFD approach the combustion processes can be predicted and the operating condition and combustion chamber optimized in existing waste to energy power plant (WTEP) or in the project phase. Fig. 1 shows the 2D engineering plan view of WTEP.

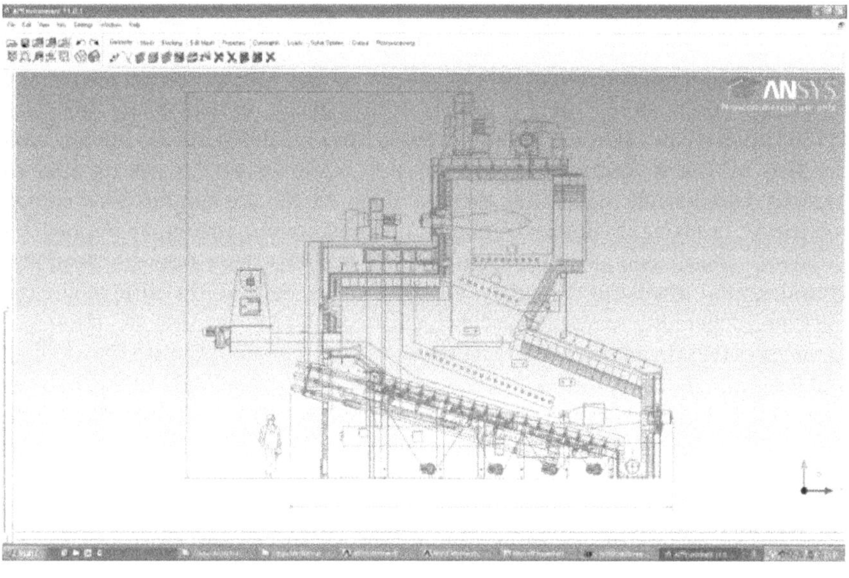

Figure 1: The 2D engineering plan view of WTEP.

On this base the WTEP was built and operates with RDF as a fuel produced with MBT from MSW. Fig. 2 shows grate, details in the primary combustion chamber with waste input (left) and the secondary combustion chamber with secondary and tertiary air inlet (right).

The 3D geometry plan on the base of engineering plans in real measure was drawn (fig. 3 – left). Each dimension was marked on the plan with corresponding input dimensions. In this way each dimension is easy and quickly modified and the entire construction can be quickly changed and redrawn.

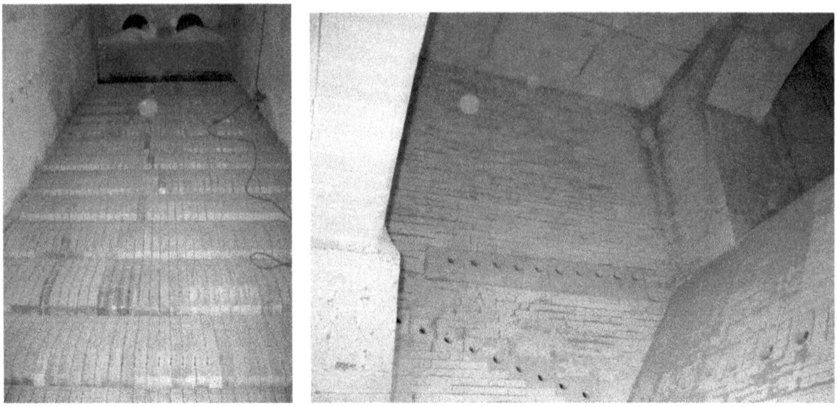

Figure 2: Grate in primary combustion chamber and secondary and tertiary air inlets.

Furthermore, the meshing of ca. 1.4 million cells (fig. 3, right) is created and the boundary conditions with entire combustion, radiation, particle tracking and other models, input and output parameters are set up. These input parameters are operating conditions like input velocities, temperatures, reactants mass flow rates (MFR), dimension value and the output parameters like temperatures, combustion products MFR and other exhaust parameters. It is known that the concentration of gaseous components which are created from MSW combustion like carbon monoxide, methane, hydrogen, nitrogen monoxide or oxygen above the bed are differently distributed along the grate set [3, 4, 6, 7, 9]. These facts are taken into consideration when the boundary conditions are defined. In this study, the measured operating parameters form existing WTEP is used as the input parameters (IP). In addition, the reactions for gaseous combustion are also defined and the entire model is solved. Many other authors have used the FLUENT solver code [3–11] but in this study it is interesting that the CFX V12.0 solver code in the Workbench 2 environment which was developed by Ansys, Inc. has been used [12, 13] and the results are quite similar.

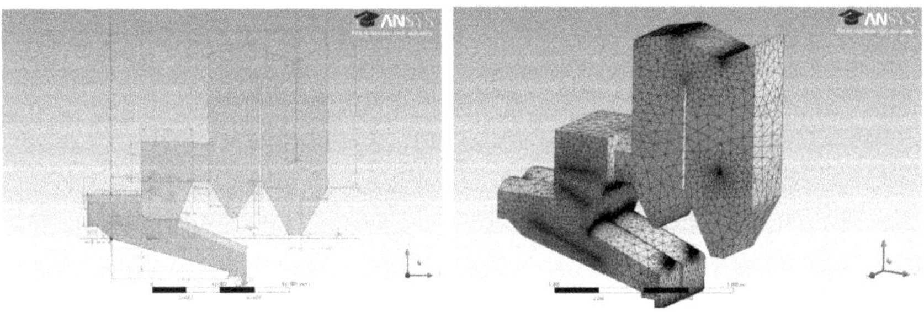

Figure 3: 3D geometry plan of WTEP with dimensions (left) and meshing (right).

The optimisation process was developed using design exploration which is a powerful tool for designing and understanding the analysis response of parts and assemblies. It is a deterministic method based on Design of Experiments (DOE) and various optimisation methods with parameters as its fundamental components. DOE techniques have one common characteristic: they try to locate the sampling points such that the space of random input parameters is explored in the most efficient way, or obtain the required information with a minimum of sampling points. Sample points in efficient locations will not only reduce the required number of sampling points, but also increase the accuracy of the response surface that is derived from the results of the sampling points. By default the deterministic method uses a central composite design, which combines one centre point, points along the axis of the input parameters and the points determined by a fractional factorial design. GDO is a constrained, multi-objective optimisation technique in which the best possible designs are obtained from a sample set given the objectives set from parameters. GDO can be used for design optimisation in three ways: the screening, the Multi-Objective Generic Algorithm (MOGA) approach and the Non-Linear Programming by Quadratic Lagrangian approach (NLPQL). The screening approach is a non-iterative direct sampling method by a quasi-random number generator based on Hammersley algorithm. MOGA approach is an iterative Multi-Objective Genetic Algorithm, which can optimize problems with continuous input parameters. NLPQL is a gradient based single objective optimizer which is based on the mathematical optimisation algorithm developed by Klaus Schittkowski. GDO has more possible objectives and they are: no-objective, minimize, maximize, and target value which is less than or equal to or greater than or equal to the input target value. The importance of the parameter is default, lower and higher. By setting these targets the three best candidates (A, B or C) are chosen which best met all the requests.

Probabilistic sensitivities are based on a statistical correlation analysis between the individual probabilistic design variables. You can review the parameters correlation data that has been used to derive sensitivities and decide if individual sensitivity values are significant or not. This information is collected in the correlation matrix of the random output parameters versus the random input variables. It can be also reviewed in the correlations that have been sampled between random input variables. The correlations between random output parameters are important if the probabilistic results of a probabilistic analysis are wanted as input for another probabilistic analysis.

When the optimisation process is finished, all results can be studied and the response charts can be visualized.

3 Results and discussion

In the study three input parameters (IP) are used: secondary air inlet velocity (InSecAirVelocity-m/s), oxygen MFR (MassInVelocity in m/s) and MWS velocity (MSWVelocity in m/s) and many other output parameters (OP): average value of following parameters at the exit of the WTEP – CO (COOut dimensionless), temperature (TempOut in Kelvin (K), CH_4 (CH4Out dimensionless), oxygen (O2Out dimensionless); average value ash travelling time along the grate (AshTime in s), average ash temperature (AshTemp in K), maximal ash temperature through the WTEP (AshMaxTemp in K), etc.

On the other hand, it has to be ensured that the maximum ash temperature does not exceed the ash melting point and flying ash deposit on heat exchangers walls, which can cause a great damage, has to be avoided. The optimisation response chart (fig. 4), which was created as a result of DOE helps the determination of the interaction among maximal ash temperature versus secondary air velocity and oxygen MFR in the secondary air inlet. The maximal ash temperature from secondary air velocity of 27 m/s to 29 m/s increases rapidly and peaked at a maximum ash temperature of 1850 K. On the other hand, there is no significant dependence of oxygen MFR in the region from 0.255 to 0.21. In this way, the damage this phenomenon can cause can be predicted and avoided.

Table 1 shows the setting and results of trade-off optimisation with IP and OP. As can be seen, three candidates, which best match the requirements, appear with belonging values. In this way the optimisation is easy and effectively used.

Fig. 5 shows the temperature field comparison by different operating condition of oxygen MFR. The temperature in secondary combustion chamber increases when enriched air is used [6, 11].

Fig. 6 shows the sensitivities among input and output parameters. It is important to know which input parameter has an influence on which output parameter and in which role (negative or positive). As can be seen, every output parameter has a huge sensitivity to the input parameter second air velocity and the output parameters have a limited sensitivity or no sensitivity to the other input parameters.

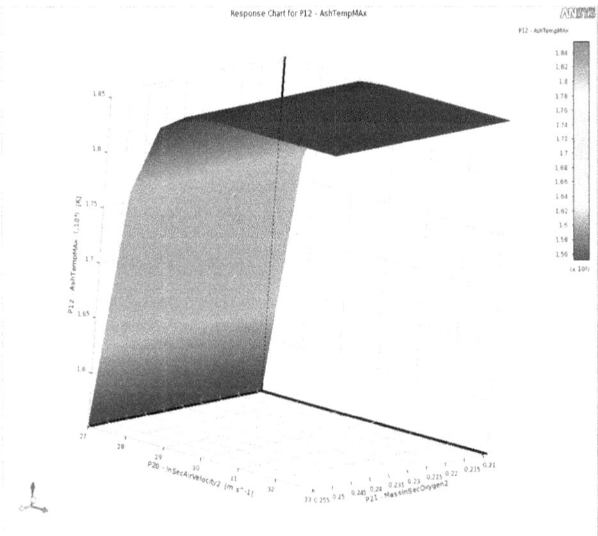

Figure 4: Response chart for maximal ash temperature vs. secondary air velocity and oxygen MFR in secondary air inlet.

Table 1: Trade-off study parameters objective settings and results.

Table of Schematic B4: Optimization

	A	B	C	D	E	F	G
1		P20 - InSecAirVelocity2 (m s^-1)	P21 - MassInSecOxygen2	P22 - MSWVelocity (m s^-1)	P5 - COOut	P6 - TempOut (K)	P7 - COSecIn
2	Optimization Study						
3	Objective	No Objective	Minimize	No Objective	Minimize	No Objective	No Objective
4	Target Value						
5	Importance	Default	Higher	Default	Higher	Default	Default
6	GDO Sample Set 1						
7	Candidate A	30,87	0,20939	0,98001	-8,0526E-06	1104,8	0,090197
8	Candidate B	32,79	0,21012	0,95367	-7,9964E-06	1104,7	0,090201
9	Candidate C	31,83	0,21084	1,0985	-8,0161E-06	1104,8	0,0902

	H	I	J	K	L	M	N
1	P8 - COSecOut	P9 - CH4Out	P10 - AshTime (s)	P11 - AshTemp (K)	P12 - AshTempMAx (K)	P13 - O2Out	P14 - COVsCo2MidRAtio
2							
3	No Objective	No Objective	No Objective	Minimize	No Objective	Maximize	No Objective
4							
5	Default	Default	Default	Higher	Default	Higher	Default
6							
7	2,0433E-05	-2,855E-06	4,9722	1174	1854,5	0,095124	0,36661
8	2,8447E-05	-2,835E-06	4,9729	1174	1854,3	0,095088	0,36663
9	2,5632E-05	-2,842E-06	4,9726	1174	1854,4	0,095101	0,36662

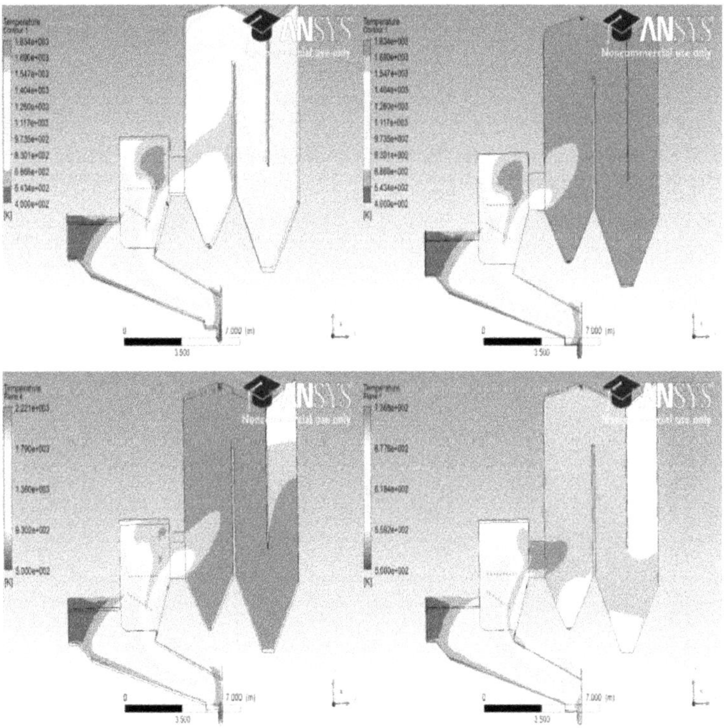

Figure 5: Temperature field by different oxygen MFR at secondary enriched air inlet.

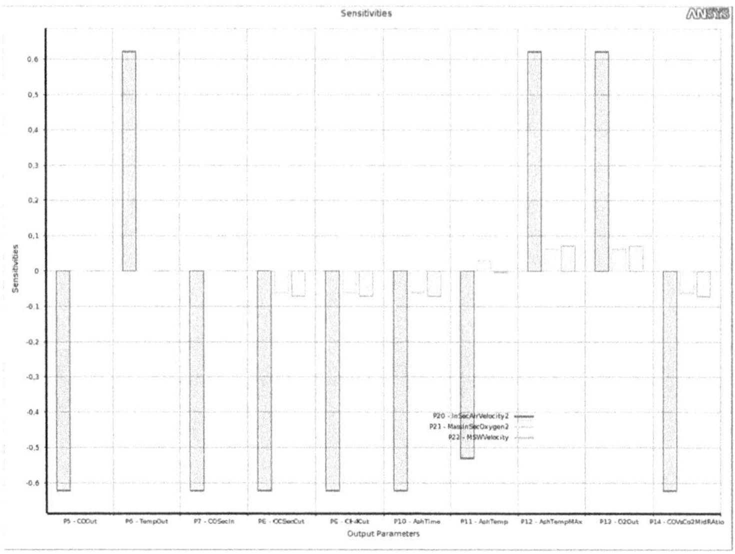

Figure 6: Local parameters sensitivities.

Fig. 7 shows the trade-off study chart for input parameter second air velocity (InSecAirVelocity2) and the output parameter CO at the exit (COOut). The feasible points to match the criteria (table 1) are shown and the dependence of input velocity is very important. For example when a velocity of 27 m/s is used, the CO MFR at the output is 0.018 and when the velocity of 27.5 m/s is used the CO MFR is only 0.0045.

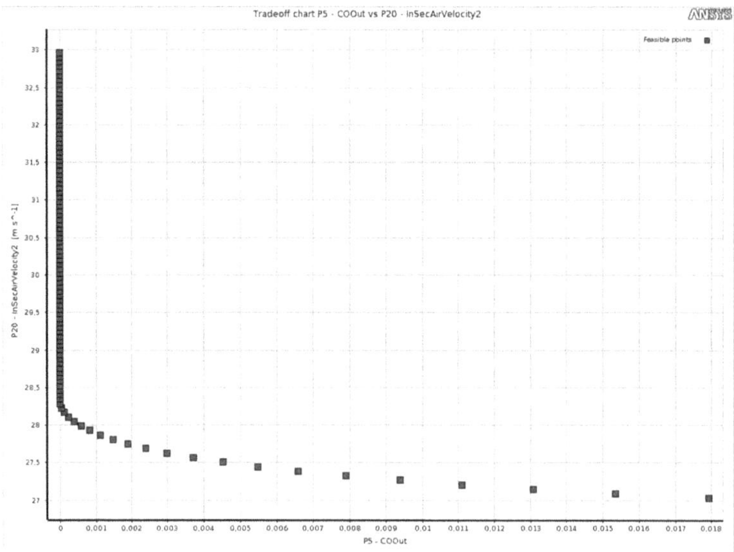

Figure 7: Trade-off chart COOut (OP) vs. second air velocity (IP).

As it can be seen, the CFD is the most convenient tool to predict the optimal conditions which have to be achieved in relation to thermal and environmental efficiency and as well as safety of the WTEP operation. With this tool, problems can also be avoided because the whole situation can be predicted with appropriate inlet boundary conditions.

4 Conclusion

The applicability of various combustion operating conditions was examined and analyzed, based on the real operating conditions. The CFD approach and the numerical optimisation was used to identify the appropriate conditions to achieve complete combustion and minimize the pollutants emission. The input and output parameters were studied and visualized in the response charts and the local sensitivities among different parameters were presented. Furthermore, the parameters correlation was identified through the parameters matrix. In the end, further simulations have to be carried out to include the geometry dimensions as input parameters.

References

[1] Yin, C., Rosendahl, L., Kaer, S.H., Clausen, S., Hvid, S.L. & Hille, T., Mathematical modeling and experimental study of biomass combustion in a thermal 108 MW grate-fired boiler. *Energy & Fuels*, **22**, pp. 1380-1390, 2008.

[2] Kokalj, F. & Samec, N., An analysis of the combustion conditions in the secondary chamber of a pilot-scale incinerator based on the computational fluid dynamics. *Journal of Mechanical Engineering*, **51**, pp. 280-303, 2005.

[3] Frey, H.-H., Peters, B., Kunsinger, H. & Vehlow, J., Characterization of municipal solid waste combustion in a grate furnace. *Waste Management*, **23**, pp. 689-701, 2003.

[4] Costa, M., Dell'Isola, M. & Massarotti, N., Numerical analysis of the thermo-fluid-dynamic field in the combustion chamber of an incinerator plant. *Energy*, **34**, pp. 2075-2086, 2009.

[5] Anderson, S.R., Kadirkamanathan, V., Chipperfield, A., Sharifi, V. & Swithenbank, J., Multi-objective optimization of operating variables in a waste incineration plant. *Computer & Chemical Engineering*, **29**, pp. 1121-1130, 2005.

[6] Yang, W., Nam, H.-S. & Choi, S., Improvement of operating conditions in waste incineration using engineering tools. *Waste Management*, **27**, pp. 604-613, 2007.

[7] Yang, Y.B. & Swithenbank, J., Mathematical modelling of particle mixing effect on the combustion of municipal solid wastes in a packet-bed furnace. *Waste Management*, **28**, pp. 1290-1300, 2008.

[8] Yang, Y.B., Sharifi, V.N. & Swithenbank, J., Converting moving-grate incineration from combustion to gasification - Numerical simulation of the burning characteristics, *Waste Management*, **27**, pp. 645-655, 2007.

[9] Yang, Y.B., Yamauchi, H., Nasserzadeh, V. & Swithenbank, J., Effects of fuel devolatilisation on the combustion of wood chips and incineration of simulated municipal solid wastes in a packed bed. *Fuel*, **82**, pp. 2205-2221, 2003.

[10] Yang, Y.B., Goh, Y.B., Zakaria, R., Nasserzadeh, V. & Swithenbank, J., Mathematical modelling of MSW incineration on a travelling bed. *Waste management*, **22**, pp. 369-380, 2002.

[11] Yu, Z., Ma, X. & Liao, Y., Mathematical modelling of combustion in a grate-fired boiler burning straw and effect of operating conditions under air- and oxygen-enriched atmospheres. *Renewable Energy*, **35**, pp. 895-903, 2010.

[12] ANSYS, Inc., Southpointe, 275 Technology drive, Canonsburg, PA 15317, United States, Software packet Workbench 2 with CFX 12.0: Help Mode.

[13] ANSYS, Inc., Southpointe, 275 Technology drive, Canonsburg, PA 15317, United States, www.ansys.com/customerportal/index.htm

A comparative study for energetic valorisation of partially digested sewage sludge

A. Alonso-Vicario[1], A. M. Macarulla-Arenaza[1],
I. Oribe-García[1] & A. Macarulla-Arenaza[2]
[1]*Deusto Institute of Technology – DeustoTech Energy, University of Deusto, Spain*
[2]*Aguas del Añarbe-Añarbeko Urak S.A., Spain*

Abstract

The common final disposition of the sewage sludge obtained after their anaerobic digestion, whose heating value is about a quarter of the coke or peat, is often used in agricultural soils or as substitute of fossil fuels in clinker kilns. In this paper, a comparative study between incineration and gasification with CO_2 as oxidant agent has been carried out. After the physical and chemical characterization of the sludge, the combustion and gasification properties have been studied by means of thermogravimetric analysis. According to the combustion profile, while the ignition temperature (200°C) and peak temperature (288°C) are of the order of any kind of plant biomass, the maximum combustion velocity is much lower, around 0.25 mg/min. Moreover, although gasification requires higher temperatures (around 790°C) for the complete transformation of the organic matter, the resulting gas, with a heating value much higher than the one contained in the dried sludge, may be used in the wastewater treatment plant, reducing both the gas natural requirements and the final sludge volume by 37%, and consequently the costs associated to the transport. The final inorganic matter may also be incorporated directly in construction materials, closing the whole loop.
Keywords: biomass, waste, sludge, energetic valorisation, co-combustion, gasification.

1 Introduction

The current global energy output based on the use of fossil fuels threatens an ecological catastrophe in the short term, on the one hand, due to their depletion through uncontrolled and irrational use today and, on the other, by exceeding the limits of the Earth's capacity to absorb its impact.

Another important source of pollution is industrial waste, which without a proper treatment is very harmful to the Earth's ecosystem. So, it is necessary to find alternative solutions that contribute to its minimization and finally its complete elimination, without endangering other social and economic factors. It is therefore imperative to find renewable sources of raw materials and to develop green technologies that address the current energy crisis and pollution problems.

The wastewater treatment is not left behind this issue. The progressive implementation of Directive 91/271/EEC, as well as the slow but steady growth of households connected to sewage treatment plants and the increased level of treated water quality has increased the amount of sludge to landfill (from 5.5 million tons of dry matter in 1992 to 9 million tons in 2005).

The use of waste to energy technologies allows the increase of flexibility of the national/regional energy systems, as well as the production of renewable energy. However, in general, the characteristics of most of these wastes are not suitable to replacing conventional fuels, so that the transformation into higher energy density fuel is necessary. The fuels thus obtained have the following advantages: lower emissions of sulfur, slag is not burning, low ash content, and contribute to improving the quality of the environment. So, biomass is converted to heat, fuels or electricity. These transformations can be divided into two groups:

- Thermo-chemical processes: Chemical transformation that takes place under high temperatures and excess oxygen (combustion) [1] , limited amounts of oxidant agents, as O_2, water or CO_2 (gasification) [2] or in lack of it (pyrolysis) [3].
- Biochemical processes: Degradation of complex molecules to simpler molecules with high energy density carried out by microorganisms, the most commonly used are alcoholic fermentation [4] and anaerobic digestion [5, 6], to produce ethanol and biogas, respectively.

The Wastewater Treatment Plant of Aguas del Añarbe, S.A. (Loyola, Spain), was designed for the treatment of urban wastewater by anaerobic digestion, so that partially recovers the energy content of sludge in the form of biogas, which is fed to a cogeneration system together with natural gas. Heat is used for the drying and dehydration phases in the sludge line and electricity for both consumption and for sale. Moreover, the final dried sludge shows low biological activity in very low volume.

However, the wastewater treatment plant produces 10 tons/day of final partially digested and dried sludge and it is expected that the production will increase to 30–40 tons/day in the near future. So, it is necessary to find alternative solutions for the biosolid whose production is expected to grow exponentially in the following years.

Among the thermochemical processes, co-combustion with fossil fuels offers economic and environmental benefits [7–9], contributing to reducing total pollutant emissions per unit of energy produced, as CO_2 emissions from biomass are not counted in the overall calculation, and also reduces both NO_x and SO_x levels on existing coal plants. Thus, biomass is usually burned in conventional power plants with steam cycles, as it was some years ago in the sugar mills [10].

In contrast, gasification is applied in more advanced and efficient systems. The gases obtained in the gasification can be used in heating processes by direct combustion or in internal combustion engines or even in gas turbines, systems that are between 10 and 15% more efficient than direct combustion of any solids and also less polluting [11], or even in more efficient systems that integrate fuel cells [12, 13].

In this paper, after a previous physical and chemical characterization of the final dried and partially digested sludge, a comparative study between combustion and gasification using carbon dioxide has been carried out by means of thermogravimetric analysis in order to demonstrate the environmental, technical and economic advantages of their use as alternative fuel.

2 Experimental work

The characterization of the biosolid has been carried out in order to know the physico-chemical properties that determine its performance as fuel by energetic valorisation. For this purpose, the following analytical methods have been used: CEA 1424, CEA 1281 and CEA 1358 standard methods for elemental analysis; drying at 105°C, and gravimetry for immediate analysis; calorimetry, atomic absorption and atomic emission spectroscopy for the determination of heavy metal and inorganic content; granulometry has been measured according to Directive 87/94/CEE; and the higher heating value following the CEA 1235 standard method.

In order to analyse microscopic properties, reactivity studies were carried out by thermogravimetry using a SETARAM Setsys Evolution thermobalance. These tests were carried out by The General Research Services of the "Instituto de Carboquímica" (ICB) of Zaragoza. Samples were heated with a temperature ramp of 10°C/min under oxidant atmosphere (with air from room temperature up to 1200°C, and 5% CO_2 in Ar from room temperature up to 900°C), followed by isotherm during 10 min.

3 Results

In general, biomass is much less dense than coal (500 kg/m^3 vs 1300 kg/m^3) [14]. It is also much more difficult to reduce to small sizes. Biomass with coal burning cannot have a particle size greater than 6.3 mm (carbon 0.1 mm) [14]. Table 1 shows the result of the fineness of the sludge, 99.9% is below 5 mm, thus ensuring its use as a supplement to coal.

Tables 2, 3 and 4 summarize the main results of the chemical characterization of the biosolid. The results of the analysis of metal content and elemental analysis are consistent with those found in the literature for this type of sludge [15, 16], with recorded values within an acceptable range to be a partially digested sludge. Thus, the content of C and H are low and its relationship (7.1) slightly lower compared to the initial (10.0) of undigested sludge. A low sulfur content (lower than in coal) translates to lower emissions of SO_x, although the nitrogen content is substantially higher, and therefore, the NO_x emissions.

Table 1: Granulometry of sludge.

Granulometry	%
> 5 mm	0.1
4–5 mm	1.0
3–4 mm	26.2
2–3 mm	60.9
1–2 mm	11.3
0.5–1 mm	0.1
< 0.5 mm	0.1
Stones and gravel in diameter > 5 mm	< 5
Impurities: metals, glass and plastic > 2 mm	< 3

In general, the biomass has a high oxygen content, even in the case of sludge despite its higher content of inorganic matter.

On the other hand, coal shows higher iron content, while plant biomass has higher potassium content than coal and sludge, but similar contents of iron, calcium, magnesium, phosphorus and chlorine as sludge. The moisture content of sludge (4.2%) is even lower than that of coal (4.8%).

Also, it is remarkable that the CaO content of sludge is higher than that of coal (5.5%). CaO is directly related to the ability of the ashes to adsorb sulfur and nitrogen by formation of the corresponding sulphates and nitrates, which is an advantage of the use of sludge as alternative fuel [17–19].

Table 2: Elemental analysis.

% dm	C	H	N	S	Cl	Ash	O(diff.)
Sludge	20.7	3.3	2.1	0.8	0.3	75.7	10.8

Table 3: Heavy metal content.

ppm	Pb	Ni	Cd	Cr	Zn	Cu	Hg
Sludge	60.1	26.5	0.8	59.0	256.5	103.9	0.4

Table 4: Analysis of the inorganic matter.

% dm	Fe_2O_3	CaO	MgO	Na_2O	K_2O	SO_3	P_2O_5
Sludge	6.6	13.7	0.9	0.3	0.9	2.1	2.3

The higher heating value of sludge is 8.8 MJ/kg, above the crust (HHV = 5.9 MJ/kg), the order of sawdust or straw (HHV = 8.4 and 12.5 MJ/kg), half those of wood or sunflower husk and about a quarter that of coke or peat.

Figs. 1 and 2 show the results of the thermogravimetric analysis with air and carbon dioxide (5% in Ar), respectively, and tables 5 and 6 show the results of the deconvolution of the previous graphs.

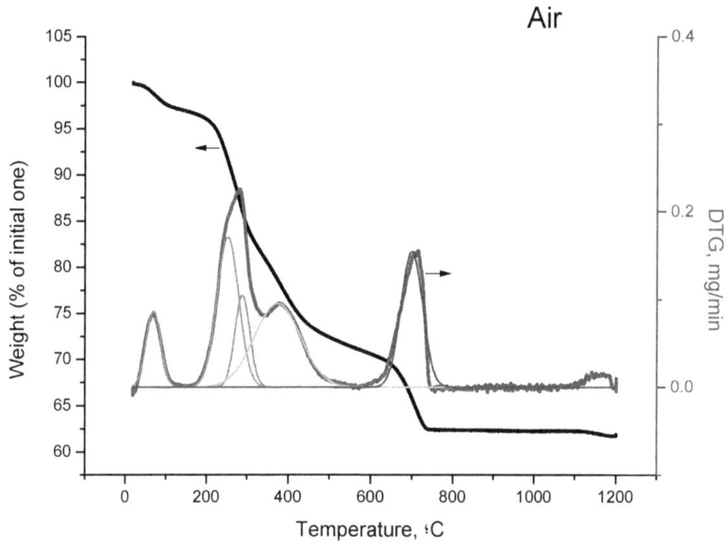

Figure 1: Thermogravimetric analysis of the combustion of the sludge.

The first peak in the profile corresponds to combustion of the sludge moisture, about 3.3%. After removing the moisture, several peaks occur as a result of the desorption of adsorbed gases.

Between 200 and 500°C there is a sudden loss of mass: about 68% of the total loss experienced during the test, which represents the formation of volatile and their ignition.

Table 5: Results of the deconvolution of the combustion graph.

Peak	T max, °C	Area, %	Loss of weight (g/100g sludge)
1	70	8.8	3.4
2	252	26.5	10.1
3	288	11.0	4.2
4	373	30.6	11.7
5	699	23.2	8.8
Total	--	*100*	*38.1*

Another peak of smaller size but not the smallest appears at very high temperatures near 700°C, probably due to partially charred residue.

The ignition temperature is the point at which the profile of combustion shows a sudden increase, in this case at 200°C, the order of any plant biomass (203°C for shells of sunflower and pine cone, 190°C for seeds rapeseed or 200°C for olive refuse [20]).

The point of combustion curve for the rate of weight loss is known as high temperature peak and is indicative of the reactivity of the sample. This temperature is 288°C, the order of the shells of sunflower (300°C), rapeseed (262°C), the pine cone (292°C), or olive refuse (264°C). The maximum combustion rate, is 0.25 mg/min, much lower than that of the aforementioned types of biomass (5.5, 2.8, 5.2, and 3.4 mg/min, respectively [20]).

As for the gasification analysis with carbon dioxide, as can be seen in fig. 2 and table 6, there are three regions or temperature ranges with loss of mass: 25–100°C (4.8%), 200–600°C (23.9%) and 700–850°C (11.6%). The first one corresponds to the loss of moisture by heating the sample. The other two are a direct consequence of the gasification, requiring at least a temperature of 790°C for a residence time long enough for the complete gasification, or 900°C and shorter times, which will affect the composition of the final synthesis gas.

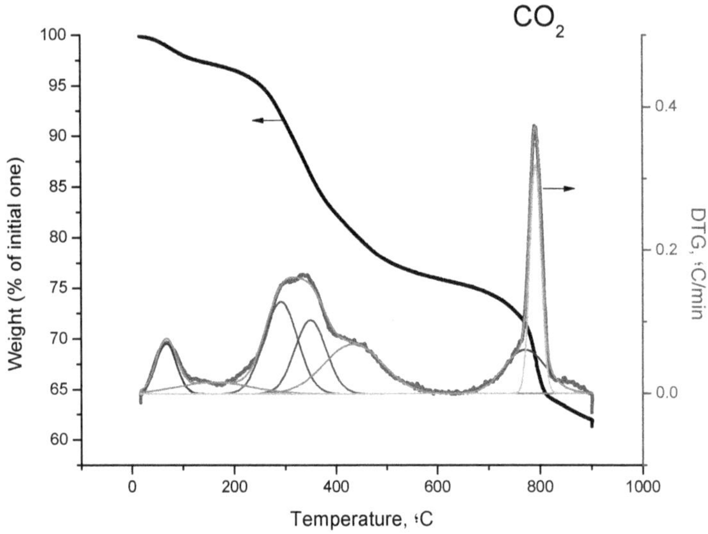

Figure 2: Thermogravimetric analysis of the activation of the sludge with CO_2.

In fig. 2 also can be observed a loss of approximately 37% of mass after the test that coincides with the total organic content of the sludge obtained in the combustion tests. This means that the gasification of the sludge results from a practically complete transformation of the organic content of the sludge to medium power combustible gas (synthesis gas is composed primarily of methane, carbon monoxide and hydrogen) with a much higher heating value (39.9 and 120 MJ/kg, for natural gas and H_2, respectively) than sludge (8.8 MJ/kg). Due to this, it would be interesting to consider it as an alternative route for energy recovery from sludge in the wastewater treatment plant or otherwise use the gasification of fossil fuels, such as power plants integrated gasification combined cycle (IGCC) generating electricity from fossil fuels. This synthesis gas obtained after the gasification is cleanly burned later in a gas turbine to produce electricity.

Table 6: Results of the deconvolution of the gasification graph.

Peak	T max, °C	Area, %	Loss of weight (g/100g sludge)
1	68	6.8	2.6
2	161	5.7	2.2
3	291	21.1	8.0
4	350	16.0	6.1
5	434	20.0	7.6
6	770	12.7	4.8
7	791	17.7	6.8
Total	--	*100*	*38.1*

The heat required for gasification of the sludge in the wastewater treatment plant itself may be obtained by combustion of biogas and CO_2 (gasifier agent) coming from the flue gases from the engine generator, which would access hot to the gasifier, reducing energy requirements. The fuel gas obtained could feed back into the gasifier to supply the heat required and the remainder to the combustion chamber of the dryer reducing the need of natural gas or co-fed with biogas in biogas generator sets. This would reduce the cost derived from consumption of natural gas, which is currently around 100,000 €/month, a proposal for maximum economic performance of the facilities that makes them viable and attractive from the point of view of exploitation.

4 Conclusions

Regardless of the proposed solutions in the coming years, dependence on fossil fuels to sustain economic development will remain so, it is necessary to find technological solutions that reduce environmental pollution due to emission of gaseous pollutants, and to seek alternative fossil fuels.

However, to ensure energy sources are affordable, safe and friendly, the environment requires exactly the same actions: energy efficiency, so no matter which of them more focus is on. From this point of view, to give an energy value to a waste, which as such is annoying, must be positive from these three points of view.

The use of sludge as alternative fuel is a net saving in CO_2 emissions because the emissions from biomass are not accounted, whereas if it is deposited in landfills, emissions due to uncontrolled fermentation itself are counted. The use of biosolids in cement kilns is accepted, but it is necessary to control specific parameters such as maximum allowable dose of sludge according to their characteristics, the range of temperature and types of fuel used, as well as the final properties of the cement. It is estimated that the use of 1 ton of sludge involves a reduction of 1.28 tons of CO_2, without including the emissions associated with transport [16].

On the other hand, the transformation of dried and partially digested sludge into synthesis gas with a higher heat of combustion is an alternative to their use in cement kilns. Also, this technology would take the total energy contained in the sludge in the sewage itself, reducing the costs associated with transport by 38% (inorganic matter cannot be enhanced through the wastewater treatment plant). The disadvantage of this route is the need for investment of the filter in the right technology, but energy-wise (HHV of the gas is well above the sludge) and environmentally (would reduce emissions from transport and the gases would be used to combustion engines in the gasification) would be most advantageous.

In 2004 the emissions of CO_2-e in Spain exceeded by 33% the expected value due, to a no real adjustment between emission reduction and economic benefit, among other factors. All the policies and advances in technology and management that lead to harmony between these two parameters will be beneficial. In particular, those that have major impact on the development of any way of recovering energy from waste and the development of sinks of CO_2-e is being expected.

Acknowledgement

This work has been possible thanks to the financial support of the BIOVALOR project (Euskadi 09 Programme – Basque Government).

References

[1] Saidur, R., Abdelaziz, E.A., Demirbas, A., Hossain, M.S. & Mekhilef, S., A review on biomass as a fuel for boilers. *Renewable and Sustainable Energy Reviews*, **15(5)**, pp. 2262-2289, 2011.

[2] Baldwin, R.M., Magrini-Bair, K.A., Nimlos, M.R., Pepiot, P., Donohoe, B.S., Hensley, J.E. & Phillips, S.D., Current research on thermochemical conversion of biomass at the National Renewable Energy Laboratory. *Applied Catalysis B: Environmental*, **115-116**, pp. 320-329, 2012.

[3] Domínguez, A., Menéndez, J.A., Inguanzo, M. & Pís, J.J., Production of bio-fuels by high temperature pyrolysis of sewage sludge using conventional and microwave heating. *Bioresource Technology*, **97(10)**, pp. 1185-1193, 2006.

[4] Shams Yazdani, S. & Gonzalez, R., Engineering *Escherichia coli* for the efficient conversion of glycerol to ethanol and co-products. *Metabolic Engineering*, **10(6)**, pp. 340-351, 2008.

[5] Iranpour, R., Palacios, R., Cox, H.H.J. & Abkian, V., Solving fecal coliform growth/reactivation in biosolids during full-scale post-digestion processes. *Water Science & Technology*, **52(1-2)**, pp. 283-288, 2005.

[6] Siles López, J.Á., Martín Santos, M., Chica Pérez, A.F. & Martín, A., Anaerobic digestion of glycerol derived from biodiesel manufacturing. *Bioresource Technology*, **100(23)**, pp. 5609-5615, 2009.

[7] Sami, M., Annamalai, K. & Wooldridge, M., Co-firing of coal and biomass fuel blends. *Progress in Energy and Combustion Science*, **27**, pp. 171-214, 2001.

[8] Hein, K.R.G. & Bemtgen, J.M., EU clean coal technology—co-combustion of coal and biomass. *Fuel Process Technology*, **54**, pp. 159-169, 1998.

[9] Harding, N.S. & Adams, B.R., Biomass as a reburning fuel: a specialized cofiring application. *Biomass Bioenergy*, **19**, pp. 429-445, 2000.

[10] Ensinas, A.V., Nebra, S.A., Lozano, M.A. & Serra, L.M., Analysis of process steam demand reduction and electricity generation in sugar and ethanol production from sugarcane. *Energy Conversion and Management*, **48**, pp. 2978-2987, 2007.

[11] Bhattacharya, A., Manna, D., Paul, B. & Datta, A., Biomass integrated gasification combined cycle power generation with supplementary biomass firing: Energy and exergy based performance analysis. *Energy*, **36**, pp. 2599-2610, 2011.

[12] Bang-Møller, C. & Rokni, M., Thermodynamic performance study of biomass gasification, solid oxide fuel cell and micro gas turbine hybrid systems. *Energy Conversion and Management*, **51**, pp. 2330-2339, 2010.

[13] Bang-Møller, C., Rokni, M. & Elmegaard, B., Exergy analysis and optimization of a biomass gasification, solid oxide fuel cell and micro gas turbine hybrid system. *Energy*, **36**, pp. 4740-4752, 2011.

[14] Demirbas, A., Sustainable cofiring of biomass with coal. *Energy Conversion and Management*, **44**, pp. 1465-1479, 2003.

[15] Fytili, D. & Zabaniotou, A., Utilization of sewage sludge in EU application of old and new methods – A review. *Renewable and Sustainable Energy Reviews*, **12**, pp. 116-140, 2008.

[16] Larrubia, M.A., Alemany, L.J., Herrera, M.C., Palacios, C., Guerrero-Prez, M.O., Ramírez, M.C. and Malpartida, I., Valorización energética de biosólidos: algunos aspectos económicos y ambientales en la EDAR Guadalhorce (Málaga). *RESIDUOS*, **17**, pp. 60, 2007.

[17] Fryda, L., Panopoulos, K., Vourliotis, P., Pavlidou, E. & Kakaras, E., Experimental investigation of fluidised bed co-combustion of meat and bone meal with coals and olive bagasse. *Fuel*, **85**, pp. 1685-1699, 2006.

[18] Suksankraisorn, K., Patumsawad, S., Vallikul, P., Fungtammasan, B. & Accary, A., Co-combustion of municipal solid waste and Thai lignite in a fluidized bed. *Energy Conversion and Management*, **45**, pp. 947-962, 2004.

[19] Nordin, A., Optimization of sulfur retention in ash when cocombusting high sulfur fuels and biomass fuels in a small pilot scale fluidized bed. *Fuel*, **74**, pp. 615-622, 1995.

[20] Demirbas, A., Combustion characteristics of different biomass fuels. *Progress in Energy and Combustion Science*, **30**, pp. 219-230, 2004.

Evaluation of thermodynamic efficiency of biowaste gasification

K. J. Ptasinski
Eindhoven University of Technology, The Netherlands

Abstract

Concerns about global climate change and depletion of fossil fuels have increased interest in biomass and other renewable energy sources. Due to shortage of clean biomass much attention is recently devoted to various biowaste streams. This paper focuses on biowaste gasification, where the produced synthesis gas can be subsequently used to produce biofuels, including Fischer-Tropsch hydrocarbons, methanol and hydrogen. The thermodynamic efficiency of biowaste gasification is evaluated for various feedstocks using energy and exergy indicators. It is shown that exergetic efficiency is lower than corresponding energetic efficiencies. Modification of biomass gasification is proposed to improve process efficiency. Finally, the conversion efficiency of overall chains biomass-to-biofuel is discussed.
Keywords: renewable energy, biomass, waste gasification, exergy analysis.

1 Introduction

With declining petroleum resources and concerns about climate change, more attention is now being directed towards sustainable production of chemicals and transportation fuels. In this respect, biomass has neutral CO_2 impact and is the only current sustainable source of carbon. The most promising biomass-to biofuel route is a two-stage process involving production of syngas from biomass gasification, followed by synthesis of transportation fuels, involving Fischer-Tropsch hydrocarbons, methanol and hydrogen. It was previously demonstrated that the gasifier is one of the least-efficient unit operation in the whole biomass-to-biofuels chain. Therefore, an analysis of the gasifier alone can substantially contribute to the efficiency improvement of the whole biofuel production technology.

The introduction of bioenergy raises many technological and also ethical questions, particularly in developing countries. In industrialized countries the agriculture is quite advanced and some countries can produce enough biomass for food as well as for biofuels. However, in some very dense populated countries, particularly The Netherlands, there is land shortage for biomass cultivation to cover all needs. On the other hand, the food production in many developing countries is rather low and these countries should first focus on producing edible biomass for food before they use biomass for biofuels. Moreover, the average conversion efficiency of sunlight into chemical energy in biomass through photosynthesis is about 0.5–1.0%, which is much lower compared to other forms of renewable energy such as photovoltaics or wind energy [1].

Therefore, the application of biomass wastes is very attractive not only from energy point of view but also for ecological reasons. The development of very efficient biomass-to-biofuels processes is currently one of the major challenges for renewable energy technology. One of the difficulties with measuring energy efficiency is the lack of consensus on the evaluation of performance of energy systems. The traditional analysis based on the First Law of Thermodynamics shows only the mass and energy flows and does not take into account how the quality of the energy and material streams degrades through the process. In this paper the exergy analysis, which is based on the Second Law of Thermodynamics is used to analyze the conversion of biomass to biofuels.

The main purpose of this paper is to study the gasification efficiency of biomass wastes. The paper starts with an explanation of evaluation of thermodynamic efficiency using the exergy concept. In section 3, the exergetic efficiency of gasification of different biomass wastes is evaluated and compared. Modification of biomass gasification to improve the efficiency is proposed in section 4. Finally, exergetic efficiency of biowaste-to-biofuel chains is presented in section 5.

2 Thermodynamic efficiency using exergy concept

The method of exergy analysis is often considered as a new method to analyze energy systems and is applied in different fields of engineering and science. The exergy concept is based on the Second Law of Thermodynamics and defined as the maximum amount of work that can be obtained from a material stream, heat stream or work interaction with respect to environmental conditions [2]. The environment is considered as a medium composed of common substances existing in abundance within the Earth's atmosphere, oceans and crust. The reference state is usually taken to be at standard temperature (T_o = 298 K) and pressure (p_o = 1 atm). Some reference species include CO_2, O_2, N_2, having a mole fraction of 0.0003, 0.2099, 0.7903 in dry air, respectively.

Exergy, contrary to the energy concept, is not conserved in a process. The sum of all entering exergy streams is always higher than that of all leaving streams and the difference is called exergy loss or irreversibility I. Irreversibility represents the internal exergy loss in a process as the loss of quality of materials and energy due to dissipation. It relates also to entropy production Π in the system according to the following equation:

$$I = T_o \Pi \tag{1}$$

Among the different forms of exergy, three forms are the major contributors to the total exergy and they are: thermal exergy, work exergy, and exergy of materials. The exergy balance of a process can be represented in the following form using exergy values of all streams entering and leaving the process:

$$\sum_{IN} E_j + E^Q + E^W = \sum_{OUT} E_k + I \tag{2}$$

where $\sum_{IN} E_j$ and $\sum_{OUT} E_k$ are exergy flows of all entering and leaving material streams, respectively, E^Q and E^W are the sums of all thermal exergy and work interactions involved in a process.

3 Exergetic efficiency of biomass gasification

Current gasifiers, such as fluidized bed and downdraft, are applied to gasify various biomass feedstocks as well as coal. The exergetic efficiency of gasification depends on the chemical composition of a fuel and in this section the efficiency of different feedstocks is compared and benchmarked against gasification of coal. These biomass feedstocks include various types of biomass and biowastes, such as wood, vegetable oil, sludge and manure, which vary greatly in chemical composition, energy content, ash and moisture content. Table 1 shows proximate and ultimate analyses for coal and the biofuels that are considered [3].

Table 1: Proximate and ultimate analysis of various biofuels.

Fuel	Proximate analysis (wt%)			Ultimate analysis (wt% of organic fraction)				
	Moisture	Ash	Organic fraction	C	H	O	N	S
Coal	11.5	8.50	80.0	78.2	4.93	13.3	1.45	1.69
Vegetable oils	0.0	0.0	100.0	75.4	11.7	12.9	0.0	0.0
Straw	12.7	6.37	80.9	48.9	5.97	43.9	0.82	0.15
Treated wood	14.6	4.41	81.0	51.5	6.03	41.3	1.22	0.09
Untreated wood	19.8	1.84	78.4	50.8	6.06	42.7	0.36	0.07
Grass/plants	24.2	5.46	70.3	49.7	6.00	42.7	1.32	0.18
Sludge	32.5	25.72	41.8	50.2	7.09	34.9	5.63	1.77
Manure	43.6	17.20	39.2	50.2	6.50	34.6	5.19	0.85

The efficiencies of gasification of these fuels are shown in fig. 1. The efficiencies are either based on energy (lower heating value LHV) or exergy (chemical and chemical and physical exergy) and are defined as the ratio of the exergy (energy, respectively) of syngas to exergy (energy, respectively) of biomass. It is assumed that biomass is gasified using air and syngas leaves the gasifier at chemical equilibrium at the gasifier temperature.

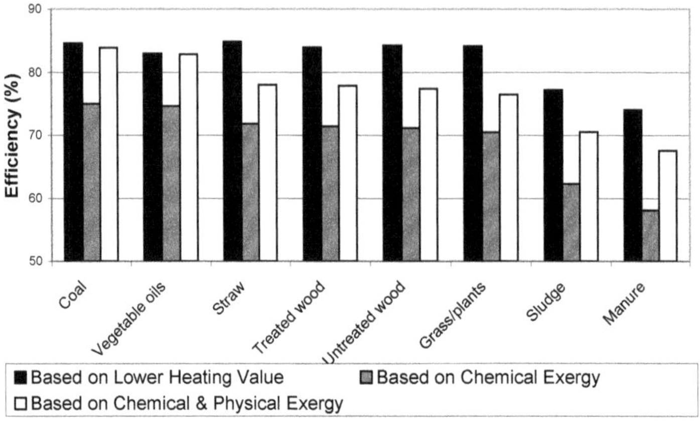

Figure 1: Gasification efficiencies of various biomass feedstocks.

The energetic efficiencies of sludge and manure gasification are much lower compared to that of other fuels, including coal. It is mainly due to high moisture content in these biowastes that results in rather low gasification temperature, about 600°C. All other fuels contain less moisture and can be gasified at much higher temperatures, in the range of 650–850°C. This figure shows that the efficiencies of coal and vegetable oil based on chemical exergy are higher that of other fuels. A similar conclusion can be drawn with respect to the efficiency based on the chemical and physical exergy. For all fuels, exergetic efficiencies are lower than energetic efficiencies; this means that less useful work (exergy) can be obtained from syngas compared to heat (LHV). The exergetic efficiencies are more meaningful than the energetic ones if syngas is used to produce transportation fuels, as the final product is work obtained from a vehicle.

The exergetic efficiency of wet biowastes, such as sludge and manure, can be improved by using the enthalpy of syngas for drying of biowastes. In this case the gasifier system consists of three units: biomass dryer operating at a temperature of 100°C placed before the gasifier, the gasifier and heat exchanger placed downstream the gasifier where the product gas is cooled down and produces the heat needed for the dryer. The temperature of the product gas leaving this gasification system is lower than the gasification temperature as heat is removed for the evaporation of water.

The gasification system is analysed for the sludge, which is dried from its initial moisture content of 32.5% to water percentages varying from 30% to 11%. This decreases the final temperature of the product gas from 600°C to temperatures between 514 and 297°C. Fig. 2 shows the effect of drying the sludge on the overall efficiency of the gasification system. It is clear that decreasing the moisture content of the biomass entering the gasifier increases all efficiencies of the process. The largest benefit is obtained when the sludge is dried from 32.5% to 19% moisture.

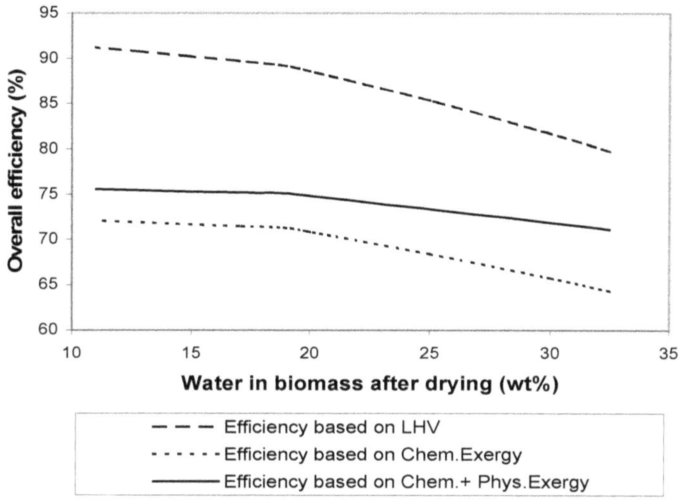

Figure 2: Improvement of gasification efficiency by pre-drying of sludge.

4 Improvement of gasification using torrefaction

The properties of biomass and biowastes can be improved by a process known as torrefaction, which comprises thermal pre-treatment in an inert gas at temperatures from 230 to 300°C [4]. The torrefied biomass has a lower O/C and H/C ratio due to the removal of water and carbon dioxide, as shown for beach wood in fig. 3, the so called van Krevelen diagram [5].

During torrefaction, the biomass partly devolatilises; this leads to a decrease of mass, but the initial energy content of the biomass is almost completely preserved in the solid product (fuel). Typically, torrefied biomass retains 70–90% of the original wood mass whereas the heating value increases by 5–25% and this way the energy density is increased by 20–40%. Torrefied biomass has a brown color and its properties are changing towards that of coal. Torrefaction is particularly attractive for a mixture of biowastes as it leads to a uniform chemical composition of the final product, e.g. in pellets.

Biomass torrefaction can be used as the pre-treatment step before biomass gasification and different systems are possible, as indicated in table 2.

Torrefaction can be coupled with the circulating fluidised bed (CFB) gasifier (IIab) where the volatiles from the torrefaction reactor are not utilized. The hot product gases from the gasifier are physically quenched with cold gas to a temperature of 800°C. Steam produced in Heat Recovery Steam Generator (HRSG) is partially used to provide heat for the torrefaction process and partially exported. The pressure of the steam generated in the HRSG is tailored for controlling the torrefaction temperature: 45 bar for torrefaction at 250°C and 90 bar for torrefaction at 300°C.

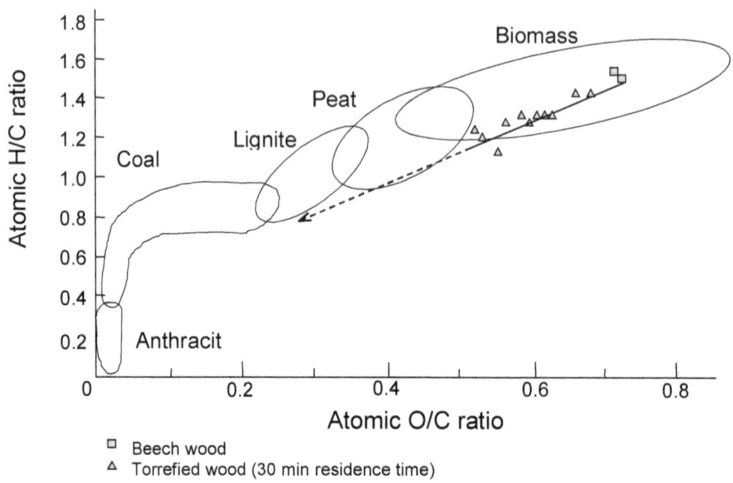

Figure 3: Change of composition of biomass in torrefaction.

In a more advanced system (see table 2 and fig. 4), wood torrefaction can be coupled with entrained flow (EF) gasifier (IIIab) [5]. In this system, the volatiles from the torrefaction reactor are introduced in the top of the gasifier to induce a "chemical quench". The conventional biomass gasification of wood (without torrefaction) in CFB gasifier (I) is used as a reference for efficiency comparison.

Table 2: Exergetic efficiency of torrefaction coupled with biomass gasification.

Efficiency (%)	Air-blown CFB gasification (950°C)	Torrefaction and air-blown CFB gasification (950°C)		Torrefaction and O_2-blown EF gasification (1200°C)	
	No torrefaction (I)	Torrefaction at 250°C (IIa)	Torrefaction at 300°C (IIb)	Torrefaction at 250°C (IIIa)	Torrefaction at 300°C (IIIb)
Gas: chemical exergy	68.6	67.2	56.2	66.8	72.4
Gas: physical exergy	1.4	1.2	1.3	0.8	1.0
Steam	6.2	5.5	4.3	4.1	2.1
Overall	76.2	73.9	61.8	71.7	75.5

Table 2 shows that the efficiency for oxygen-blown EF gasification of torrefied wood (IIIb) is quite comparable to that of air-blown CFB gasification of wood (I), and slightly lower than that if torrefied wood is produced at 250°C (case IIIa). However, the highest amount of chemical exergy retained in the product gas is observed for case IIIb; this is very convenient for the subsequent chemical synthesis of transportation biofuels.

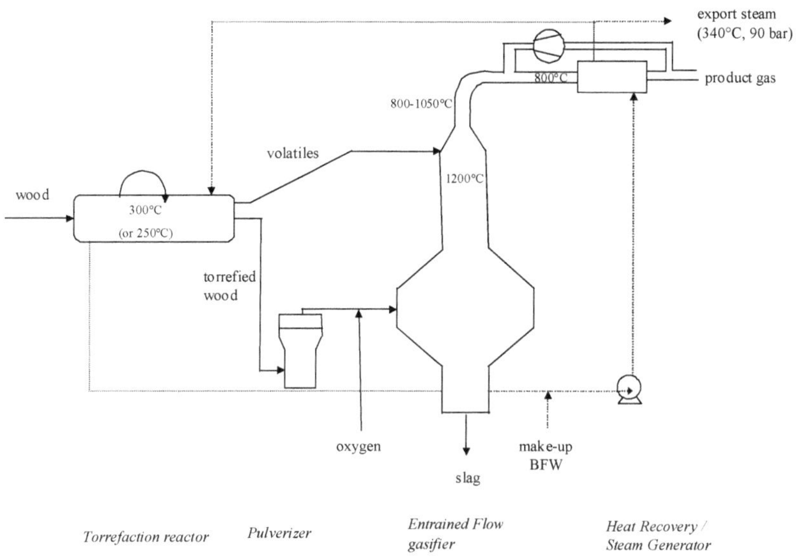

Figure 4: Torrefaction coupled with oxygen-blown EF gasifier.

5 Exergetic efficiency of biofuels from biomass gasification

Currently the most promising route to sustainable transportation fuels is based on production of syngas from biomass gasification, followed by synthesis of biofuels [6]. Production of potential transportation fuels, including Fischer-Tropsch hydrocarbons, methanol and hydrogen from various biomass feedstocks and biowastes is attractive using this method [7].

Fig. 5 shows a schematic diagram of conversion process for production of biofuels from biomass gasification. A wide range of biomass sources, such as traditional agricultural crops, residues from agriculture and foresting, sewage sludge and municipal solid wastes can be used to make biofuels. These biomass feedstocks vary greatly in chemical composition, energy content, ash and moisture content and the best and economically attractive feedstock can be selected.

For wet biomass feedstocks, drying is needed to adjust the moisture content into a desired value. In the gasifier, biomass reacts at high temperature (typically 900°C) with air, oxygen or/and steam to produce syngas, which contains CO, H_2, CO_2, CH_4, H_2O and N_2. CO and H_2 are the building reagents for synthesis of different biofuels. Subsequently syngas is cleaned from undesired components produced in the gasifier, such as tars, H_2S, NH_3 and HCl. The synthesis of most biofuels is usually a high-pressure process and therefore syngas is compressed up to the desired reactor pressure. Before the biofuel synthesis the ratio between CO and H_2 in syngas is adjusted in a water gas shift (WGS) reactor, according to the chemical reaction:

$$CO + H_2O = H_2 + CO_2 \tag{3}$$

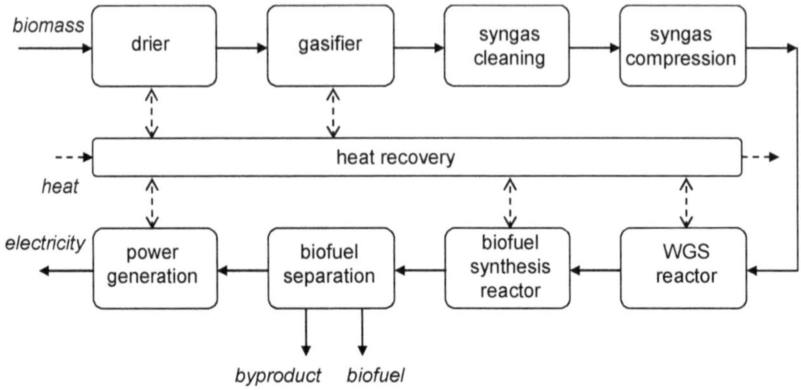

Figure 5: Schematics of biomass-to-biofuels production.

In a catalytic synthesis reactor specific biofuels, such as Fischer-Tropsch hydrocarbons or methanol are produced. In case of hydrogen production, only WGS reactors are sufficient and no specific synthesis reactors are needed. The biofuel formed in the synthesis reactor is separated from by-products in a separation unit. The remaining by-products contain usually high energy content and they can be used to produce electricity in a power generation unit or process steam in the heat integration unit. The heat recovery is particularly needed as in the overall process synthetic gas is cooled from a high temperature in the gasifier to much lower temperature in the biofuel synthesis reactor.

The process illustrated in fig. 5 can be used to produce Fischer-Tropsch (FT) hydrocarbons from sawdust. In this process 57 ton/h of sawdust is converted into 5.6 ton/h of liquid FT products and 4.0 MW of electricity is also produced. On weight basis the net output of hydrocarbon fuel is only 15.2%. The exergy analysis shows that the biomass feed contains 210.8 MW exergy and the final liquid products contain 72.9 MW with co-production of 4.0 MW of electricity.

Methanol may play a significant role as a synthetic fuel for the future as it has higher energy content per volume than the other alternative fuels. Generally, methanol can be produced from any organic source including biomass and municipal wastes according to the flowsheet shown in fig. 5. Using sewage sludge as a feedstock (50,000 tons dry solids/year), the plant can produce 4300 tons methanol/year. On weight basis the net output of methanol is only 8.6% but on the exergy basis 22.8 MW of sludge exergy is converted into 15.4 MW of methanol exergy.

Various biomass resources, such as wood, vegetable oil and manure can be also used to produce hydrogen. Currently, hydrogen is the most promising energy carrier for the future. Using wood as a feedstock the plant converts 1000 kg/h biomass into 76 kg/h hydrogen and 1330 kg/h CO_2-rich gas as the by-product. The net output of hydrogen on mass basis is thus 7.6% only. The exergy analysis indicates that 4.64 MW of wood exergy is transformed into 2.54 MW exergy of hydrogen and 0.61 MW of CO_2-rich gas.

Fig. 6 shows the exergetic efficiency of various biomass-to-biofuel routes that are discussed above but also including hydrogen produced from biomass using biochemical processes such as fermentation and anaerobic digestion. The exergetic efficiency for hydrogen-from-natural-gas process by steam methane reforming (SMR) is also reported in this figure, as a typical value for fossil fuels [8]. All reported biomass routes have lower exergetic efficiency compared to fossil fuels. The highest efficiency is for hydrogen-from-biomass processes from high quality feedstocks, such as wood or vegetable oil. It is demonstrated that biomass gasification is the least-efficient process step in the overall biomass-to-biofuels route.

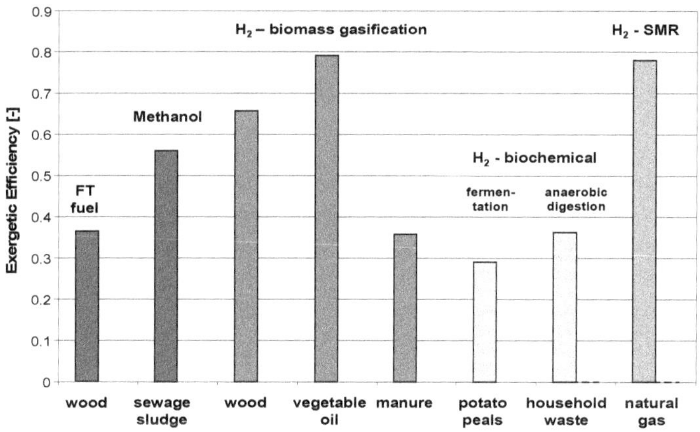

Figure 6: Exergetic efficiency of various biomass-to-biofuels processes.

6 Conclusion

Biomass as a gasifier fuel can provide the world more renewable transportation fuels. However, major limitations are shortage of land and water and competition with food production. Therefore application of biowastes is very attractive not only from energy point of view but also for ecological reasons.

The most promising large-scale production of biofuels consists of gasification of biomass (biowastes) and subsequent conversion into liquid (gaseous) fuels, such as FT hydrocarbons, methanol, and hydrogen. In all investigated processes the largest exergy losses are due to biomass gasification, which are intrinsic. Wet wastes (e.g. sludge, manure) are attractive feedstocks in many countries due to limited availability of so-called clean biomass. However, exergetic efficiency of conversion of wet biowastes is quite low as gasification at optimum conditions is not always possible.

The efficiency of gasification of wet biowastes can be improved by using the enthalpy of the syngas for drying of biomass. The properties of biomass, even of dry biowastes, can be improved by a thermal pre-treatment called torrefaction. In this process, the O/C ratio in biomass is reduced, in addition to the higher energy density; this results in the improved thermodynamic performance of gasification.

References

[1] Herman, W.A., Quantifying global exergy resources. *Energy,* **31**, pp. 1349–1366, 2006.

[2] Szargut, J., Morris, D.R. & Steward, F.R., *Exergy Analysis of Thermal, Chemical and Metallurgical Processes*, Springer Verlag: Berlin, 1988.

[3] Ptasinski, K.J., Prins, M.J. & Pierik, A., Exergetic evaluation of biomass gasification. *Energy,* **32**, pp. 568–574, 2007.

[4] Bourgois, J. & Guyonnet, R., Characterization and analysis of torrefied wood. *Wood Science Technology*, **22**, pp. 143–155, 1988.

[5] Prins, M.J., Ptasinski, K.J. & Janssen, F.J.J.G., More efficient biomass gasification via torrefaction. *Energy*, **31**, pp. 3458–3470, 2006.

[6] Huber, G.W., Iborra, S. & Corma, A., Synthesis of transportation fuels from biomass: chemistry, catalysts, and engineering. *Chemical Reviews,* **106**, pp. 4044–4098, 2006.

[7] Ptasinski, K.J., Efficiency analysis of production of modern energy carriers from renewable resources and wastes. *Ecosystems and Sustainable Development VI*, eds. E. Tiezzi, J.C. Marques, C.A. Brebbia & S.E. Jørgensen, WIT Press: Southampton, Boston, pp. 239–249, 2007.

[8] Rosen, M.A. & Scott, D.S., Comparative efficiency assessments for a range of hydrogen production processes. *International Journal of Hydrogen Energy*, **23**, pp. 653–659, 1998.

Batch waste gasification technology: characteristics and perspectives

P. De Filippis[1], M. Scarsella[1], N. Verdone[1] & G. B. Carnevale[2]
[1]*Chemical Engineering Department, Sapienza University of Rome, Italy*
[2]*Energo s.r.l., Italy*

Abstract

The problem of municipal solid waste (MSW) disposal is expanding rapidly in all countries as a consequence of the habits and lifestyles of modern consumers that cause a continuous increasing volume of waste. To date, landfills are still the primary destination of waste. However, with landfill tipping fees rising and their proven negative environmental impacts, cleaner and less costly alternatives for municipal waste disposal should be identified and implemented. The energy recovery from MSW, known as "waste-to-energy" (WTE), is one of such alternatives. Waste-to-energy reduces the amount of materials sent to landfills, prevents air/water contamination, improves recycling rates and lessens the dependence on fossil fuels for power generation. The centrality of thermal processes in the WTE practice makes highly necessary the evaluation of the problems associated with both the release and formation of contaminants and the degree of pre-processing required in conversion of MSW into a suitable feed material. Among the currently WTE technologies, one that is finding some interest in Italy is the so called "Molecular Dissociation". This technology developed by Energo is claimed to have some advantages, the main of which are the possibility to use as feed material unsorted MSW, the low emission level and, particularly, the high flexibility due to its modular design that makes the process well adaptable to the seasonal changes and population growth. Experimentation conducted in a pilot plant greatly contributed to understanding of the behaviour of the process. On the basis of the obtained experimental results, the paper illustrates advantages and disadvantages of this technology.
Keywords: waste gasification, thermal processes, batch processes, MSW.

1 Introduction

Accelerated economic development of many regions of the world in recent decades and the changes in the consumption patterns of modern lifestyle have resulted in a continuously increasing volume of waste, exasperating the problems related with municipal solid waste (MSW) disposal.

Notwithstanding the growing difficulties of many cities to find sites where to dispose the large quantities of MSW produced daily, to date landfills are still the primary destination of waste worldwide. However, landfill tipping fees rising and their proven negative environmental impact have substantially changed the processing of MSW, improving the landfill efficiency and implementing cleaner and less costly alternatives for MSW disposal.

MSW processing is, in fact, now focused on reducing the amount of materials going into the landfill using an integrated strategy that involves various actors of the waste chain. This strategy begins with source reduction modifying how products are made, and continues increasing material recycling by converting it into reusable products and treating the waste through thermal processes before entering the landfill. In this strategy thermal processing remains a functional piece of the MSW puzzle because of the ability to reduce feedstock by an average of 90% in volume and 75% in weight and because these processes can help the cost balance of MSW management [1].

Thermal processes are also at the base of the so called "waste-to-energy" (WTE) practice that besides the reduction of the amount of materials sent to landfill, prevents air/water contamination, improves recycling rates and lessens the dependence on fossil fuels for power generation. The two most commercially available forms of large scale WTE are combustion and gasification.

Combustion is a well-established practice, largely used in all countries around the world. In Europe with more than 50 million tons treated each year, it accounts for energy of about 20 million of MWh [2]. However, during combustion, dioxins and furans can form in the flue gas as they leave the combustion chamber and cool to 600–300°C. These dioxins and furans, although captured before flue gas leaves the plant, are the primary catalyst for political and environmental opposition to the expansion of the WTE industry.

Alternative to combustion is gasification. This technology is still in its early stages as a large commercial industry, but it is claimed to represent the next evolution in the thermal processing of MSW.

A technology in the middle between gasification and combustion is that developed by Energo with the trade mark name of "Molecular Dissociation". This WTE technology, is now finding some particular interest in Italy. The objective of this paper is to critically analyze this technology also considering the results of experimentation conducted in a pilot plant.

2 MSW thermal treatments

It is generally accepted that MSW represents a material very difficult to handle, segregate and feed in a controlled manner to a waste-to-energy facility. Moreover

MSW has a high tendency to form fused ash deposits on the internal surfaces of furnaces and high temperature reactors and to form bonded fouling deposits on heat exchanger surfaces.

The products of the thermal treatments of MSW are also very aggressive, the flue gases are erosive and, depending on the considered process, can contain tars, heavy metals, halogens and alkaline compounds responsible of environmental and operational problems. Furthermore, in the case of combustion, the relatively high levels of chloride containing species in the flue gases can lead to, other than the formation of dioxins and furans, high rates of metal wastage of heat exchange tube surfaces due to high temperature corrosion.

As a consequence, the key to achieve cost efficient, clean energy recovery from municipal solid waste will be in evaluating the problems associated with both the release and formation of contaminants and the degree of pre-processing required in conversion of MSW into a suitable feed material.

Unsorted MSW is not suitable for most thermal technologies because of its varying composition and size of some of the constituent materials, and owing to the presence of undesirable materials which can negatively affect the process or the emission control systems.

The purpose of pre-processing is to produce a feed material with consistent physical characteristics and chemical properties. Pre-processing operations are also designed to produce a material that can be safely handled, transported and stored.

2.1 The MSW gasification process

The increasing technical, environmental and public dissatisfaction with the performance of conventional combustion (incineration) processes is the main reason for the growing popularity of gasification processes for the treatment of solid wastes [3]. There are numerous solid waste gasification facilities operating or under construction around the world.

Gasification is a process that devolatilizes solid or liquid hydrocarbons and converts them into a low or medium calorific value gas. Two processes must take place in order to produce a useable fuel gas. In the first stage, pyrolysis releases the volatile components of the fuel at temperatures below 600°C. The by-product of pyrolysis that is not vaporized is called char and consists mainly of fixed carbon and ash. In the second gasification stage, the carbon remaining after pyrolysis is either reacted with steam or hydrogen or combusted with air or pure oxygen.

Gasification has several advantages over traditional combustion of MSW. It takes place in a low oxygen environment that limits the formation of dioxins and of SOx and NOx. Furthermore, it requires just a fraction of the stoichiometric amount of oxygen necessary for combustion. As a result, the volume of process gas is low, requiring smaller and less expensive gas cleaning equipment. The lower gas volume also means a higher partial pressure of contaminants in the off-gas, which make easier the complete adsorption and the particulate capture. Finally, gasification generates a fuel gas that can be integrated with combined cycle turbines, reciprocating engines and, potentially, with fuel cells that convert fuel energy to electricity more efficiently than conventional steam boilers.

3 The batch gasification systems

Although thermal processes are generally continuous, a certain interest is actually emerging for batch processes.

The two batch gasification systems that are earning some commercial success are the “Brookes Gasification Process” (BGP) and the “Batch Oxidation System” or “Molecular Dissociation”.

The BGP technology is based on a fixed-bed, indirectly-heated, batch-fed gasifier, suitable for processing a wide range of feedstock materials, including swine manure solids. The BGP gasifier, more correctly seen as a pyrolyzer, consists of two chambers: a burning chamber and a ‘gasification’ chamber. Hot gases from the burning chamber heat the refractory bricks that line each of the two chambers, while the feedstock (e.g., separated swine solids) sits undisturbed on the hearth of the gasification chamber where it is indirectly heated to 600–1000°C by heat transfer [4].

In this way, the feedstock is broken down into low molecular weight gases and volatile hydrocarbons, which are subsequently drawn into the burner chamber. Energy from these out-gases provides fuel that can sustain the entire gasification process. This process, initially designed as a cremator, evolved into an efficient system for the safe disposal of biohazard waste. However, it does not permit energy recovery and for its construction characteristics is not suitable for treating the large amount of MSW produced daily, even from a small town.

The “Molecular Dissociation” system, in contrast, was specifically designed for the thermal treatment of small amounts of untreated waste. Also, in this technology, the process is conducted using two stages (chambers): a batch gasification stage and a batch or continuous oxidation stage. The gasification occurs as a result of a partial oxidation process using substoichiometric combustion. The untreated solid waste is initially loaded into a primary chamber where it is thermally reacted under starved-air conditions and transformed into burnable gases and ash. Unlike typical thermal treatment methods, the gasification reactions occur at relatively low temperatures (450–550°C) under quiescent conditions. The process is initiated by an ignition burner, situated towards the top of the gasification chamber. Once the gasification has commenced, this burner is turned off and the thermal process continues, controlled by the input of a small amount of air from the base of the chamber. The low gasification temperature minimizes the production of airborne 'fly ash' particulates, carryover of toxic metals and NOx. The gasification process ensures nearly 100% burn-out of the waste and the by-product white ash with minimal residual carbon. Metals and glass in the waste remain unmelted with the ash in inert form and can be recovered by conventional recycling methods.

The combustible gases produced in the gasification chamber enter the secondary combustion chamber where they are mixed with further air (secondary air) and oxidized at much higher temperatures (850–1200°C) which ensure the complete combustion. The gases in general can support this process; however a secondary burner ensures that minimum temperatures are maintained.

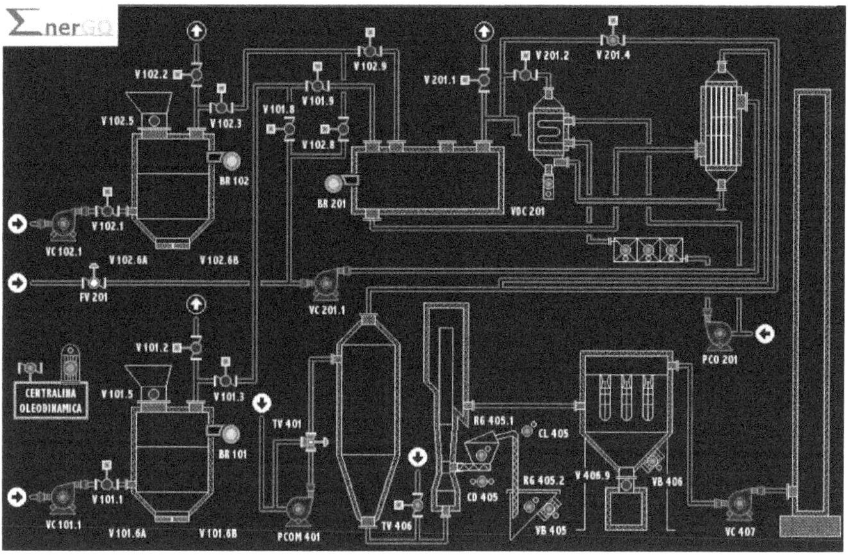

Figure 1: General scheme of the pilot plant.

Figure 2: Picture of the pilot plant.

The energy from hot gas effluent can be recovered and used to produce steam and electricity.

A "Molecular Dissociation" pilot plant designed and constructed by Energo is sited in Peccioli inside the Belvedere SpA landfill. The plant consist of two gasification chambers each with a capacity of 8 m^3, an oxidation chamber designed to assure a permanence time of at least two seconds at a temperature higher than 850°C, in accordance with the Italian legislation, and a flue gas treatment system.

The general scheme is reported in fig. 1, while fig. 2 reports a picture of the pilot plant with the two gasification chambers in the foreground.

The plant works with unsorted MSW, whose main physico-chemical characteristics are reported in table 1.

Table 1: Main physico-chemical characteristics of the charged MSW.

Parameter		Metal analysis	mg/kg
Low heating value (MJ/kg)	11.8	Al	1881.23
Water (% wt)	15.4	As	1.00
Ash (% dry wt)	12.6	Ba	65.36
		Cd	1.03
Elemental analysis		Cr	25.76
(dry basis)		Fe	3197.05
C (%wt)	50.42	Mg	83.03
O (%wt)	27.87	Hg	0.10
H (%wt)	5.85	Ni	12.11
N (%wt)	1.69	Pb	57.10
Cl (%wt)	1.35	Cu	25.09
S (%wt)	0.23	V	7.67
F (mg/kg)	73.8	Zn	95.39

The ash discharged at the end of the process contains low amounts of unburned carbon (typically less than 3%), and high amounts of metals. The main metals composition of the discharged ashes is reported in table 2.

Table 2: Metals composition of the discharged ashes.

Metal analysis	mg/kg
Al	14310
As	4.25
Ba	629.3
Cd	3.7
Cr	335.17
Fe	13596
Mg	429.3
Hg	-
Ni	119.10
Pb	268.27
Cu	958.4
V	26.34
Zn	2962.9

The gas composition at the exit of the oxidation chamber before entering the gas treatment system is characterized by a low amount of total particulate, generally less than 10 mg/Nm^3 and by a very low content of metals.

The quite total recovery of metals in the ash is compatible with the low temperature reached in the gasification chamber.

4 Critical analysis of the "Molecular Dissociation"

The experimental results confirm the interesting characteristics of the studied process that thanks to the combination of low gasification temperature and almost quiescent conditions allows one to minimize the problems associated with the presence in the off-gas of particulates and toxic volatile metals, responsible of fouling deposits on heat exchanger surfaces.

Furthermore, owing to the modularity of the system, which allows the change of the treatment capacity of the plant, the "Molecular Dissociation" process appears potentially ideal for a municipality with very high seasonal shift of population, and consequently with a seasonal variation of MSW production. This is particularly true in the islands subjected to a high touristic flux, where it can reduce the need for the transportation of the waste stream generated by the community. This plant flexibility can also potentially satisfy the waste disposal necessity of seasonal industry.

Nevertheless, some of the most attractive characteristics of this technology represent also its limits. First of all, due to the low working temperature, the oxidation kinetic is very slow, generally more than an order of magnitude with respect to conventional thermal treatments. As a consequence, larger plants are needed to treat the same amount of waste. Furthermore, due to the intrinsic nature of the batch technology, it is impossible to maintain a constant gas quality that will follow periodic variation during the process, being more energetic at the beginning of gasification and progressively less energetic with the prosecution of the process. This implies that the gas produced from the primary chamber is not suitable as syngas, but for an efficient energy recovery it is necessary to burn this gas directly in a secondary chamber taking advantage of both its sensible and latent heat content. Also in this manner, to optimize the energy recovery, it is necessary to manage the process working with more batch gasification chambers, each one in a different reaction advancement state, as reported in fig. 3.

This reduces the flexibility and the advantages of the modularity of the process. Moreover, the use of low capacity plants constitutes a handicap for the energy recovery because of the high amount of energy lost during the start up and the cooling down of the chambers at the end of the process.

5 Conclusions

The batch gasification process could represent a viable alternative for small community waste treatment allowing the reduction of emissions and cost due to waste transportation. The low working temperature of the process assures the minimization of the problems associated with the presence in the off-gas of

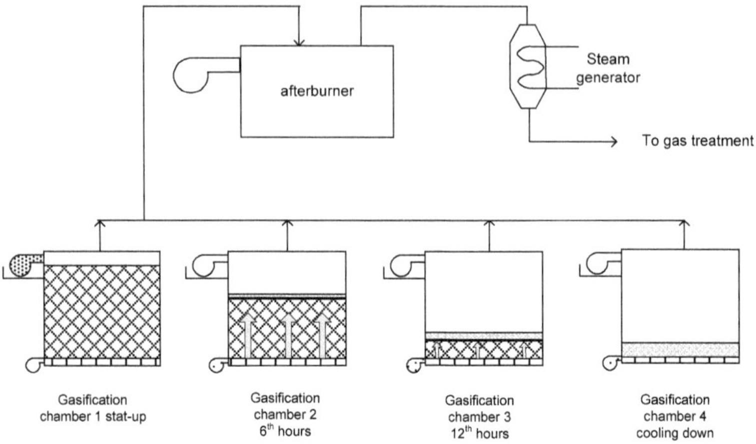

Figure 3: Typical process arrangement.

particulates and toxic volatile metals, responsible for fouling deposits on heat exchanger surfaces. Furthermore, owing to the mild conditions, ashes can represent a source of secondary raw materials.

However, the low working temperature represents also the main limit of this batch technology: it has, in fact, to be considered that the consequent slow reaction kinetics penalizes the capacity of the process and makes difficult the use of high efficiency energy recovery systems.

References

[1] Schilli, J.W., Using gasification to process municipal solid waste. *HDR Innovations*, **12(4)**, 2004.

[2] ISWA Working Group on Thermal Treatment of Waste, *Energy from Waste: State-of-the-art-report*, 5th ed., Rambøll: Denmark, 2006.

[3] Malcow, T., Novel and innovative pyrolysis and gasification technologies for energy efficient and environmentally sound MSW disposal. *Waste Management*, **24**, pp 53-79, 2004.

[4] Waste management programs, NC State University, *Development of Environmentally Superior Technologies, Phase 2 Report*, July 2005. Online. http://www.cals.ncsu.edu/waste_mgt/smithfield_projects/smithfieldsite.htm

Experimental study and model validation of waste gasification in an up-draft fixed-bed gasifier

N. Verdone, P. De Filippis, M. Scarsella & B. de Caprariis
Sapienza University of Rome, Italy

Abstract

Gasification has been identified as a key technology to enhance the environmental tolerability of low quality fuels such as waste and biomass. In this work, the performances of a laboratory scale gasification process fed with waste are reported. Among the several technical choices, the up-draft fixed-bed gasifier was selected as an interesting solution for heat generation in small-scale applications, due to the characteristics of simple geometry and low cost. The experimental setup is composed of an up-draft gasifier followed by a reactor used as filter to remove the particulate and as second thermal and catalytic stage to convert the produced tar to lighter species. A literature model has been adapted to the case under study to analyse the influence of operative parameters such as oxidant flow rate (equivalent and air/steam ratio values) and gasification temperature of the process. The original literature model considers the species gas evolution along the axial coordinate only and does not include time dependency. To make the model time dependent, the consumption time of the gasification fuel bed estimated from experiments was introduced. Furthermore the modelling of the oxidation zone was introduced, adding the char combustion equation. Since the model concerns only the gasification, the up-draft process was split into two consecutive steps to allow direct comparison between experimental and simulated data: first the drying and pyrolysis processes and then the fixed bed gasification. During the pyrolysis, a second stage reactor for the tar reforming was included, allowing a tar reduction of about 85 %. The model was successfully validated with experimental data and then was used to predict the operative parameters that determine the optimal syngas composition. The best syngas composition (35% CO and 10% H_2) was obtained with an equivalent ratio of 0.6 and a bed temperature of 1100 K.
Keywords: auto fluff, disposal, gasification, energy recovery.

1 Introduction

Automotive shredder residue (ASR) or auto fluff is generated by technologies that recover non-ferrous metals and plastics from end-of-life vehicles, appliances and other light scrap. The main flow of ASRs is generated by the shredding of end-of-life vehicles (ELVs). The ASR is principally composed of textiles (25%), plastic (23%), sponge (17%), as well as sand and solids (16%) [1]. The ASR represents 20% of the ELVs that is not recycled and is disposed in landfill in the majority of the EU countries [2]. Sometimes, the presence of heavy metals and chloride from metals and PVC respectively makes ASR a hazardous waste causing increasing costs for its disposal.

The incineration is another possibility for ASRs disposal, but the high content of inert leads to a low grade heating value making difficult the combustion of the ASR alone. Usually the co-combustion with municipal solid waste is preferred.

The quantity of ASR is forecast to rise in the coming years because of the increasing amount of polymers replacing metals in new vehicles. ASR valorisation treatments are necessary according to the limit of 13 MJ/kg stated for waste disposal by the European Directive 1999/31/EC and according to the Directive 2000/53/EC that claims for ELVs at least 85% recycling rate and 95% recovery rate by the year 2015, thus allowing the disposal in landfill for maximum 5% ELV total weight [3]. Thus energy recovery from ASR will become one of the most attractive possibilities to dispose auto fluff.

In this work, it is proposed to recover energy from auto fluff by gasification. In the gasification process, a combustible gas from a solid fuel with low grade heating value is produced. The up-draft fixed-bed gasifier was selected as an interesting solution for heat generation due to the characteristics of simple geometry and low cost.

A one-dimensional model was implemented to simulate the behaviour of the gasification stage. The composition of the produced gas was estimated and compared with experimental data showing a good agreement. The model was used to optimize the operative conditions of the gasifier in order to obtain the best syngas composition.

2 Experimental set-up

2.1 Experimental apparatus

A laboratory-scale gasification plant was designed and constructed. Its main components are schematically represented in fig. 1. The plant is composed of an updraft gasifier followed by a catalytic reforming reactor, where the tar cracking occurs and the dragged particulate gasification reactions are completed. The catalyst used in the reforming reactor is made of aluminium oxide. The two reactors are cylindrical steel tubes of 4 cm diameter (*D*) and 60 cm length (*L*). The high value of the *D*/*L* reactor ratio makes it difficult to have an autothermal process, therefore the two reactors are heated by external tubular ovens. A grate, placed at 10 cm from the bottom of the gasifier, is used as support for the solid combustible and as air distributor. The fuel can be injected continuously through a cochlea whereas the ash removal is discontinuous. The two gasification agents, steam and air, are premixed, preheated and then injected into the reactor in counter-current with respect to the solid flow.

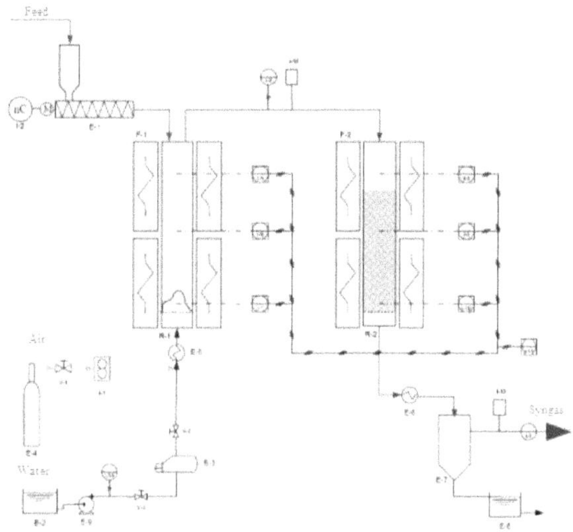

Figure 1: Laboratory scale plant.

Temperature profiles along the gasifier axis are measured by four thermocouples placed in a ceramic protective tube. The CO, CO_2 and CH_4 concentrations are monitored continuously. The H_2 concentration is measured with a gas chromatograph every 15 min. The plant is equipped with two gas outflows: at the exit of the gasifier and at the exit of the reforming reactor. A condenser for the collection of the vapours produced during the pyrolysis is placed after the reforming reactor.

2.2 Materials

ASR is a heterogeneous material composed of a mixture of plastics, rubber, resins and textiles and by a small amount of glass fragment and unrecovered metals. Because of the heterogeneity of this material, samples were first accurately grinded and mixed. The immediate and elemental analyses were performed and the calorific value was calculated experimentally with a Mahler bomb calorimeter. Average values of these properties are reported in table 1.

3 Model

A literature model [4] for a down-draft gasifier was adapted to our case. The model is one-dimensional, the bed and gas properties variations were considered negligible and the gases were assumed to behave ideally.

Five reactions were taken into consideration [5]:

1) $C + O_2 \rightarrow CO_2$
2) $C + CO_2 \rightarrow 2CO$
3) $C + H_2O \rightarrow CO + H_2$
4) $C + 2H_2 \rightarrow CH_4$
5) $CH_4 + H_2O \rightarrow CO + 3H_2$

Table 1: ASRs properties.

Immediate Analysis	Value
Humidity (% mass)	5%
Volatile matter (% mass dry basis)	32%
Fixed carbon (% mass dry basis)	11%
Ash (% mass dry basis)	57%
Upper calorific value (MJ/kg)	16.2
Elemental Analysis	(% mass dry basis)
C	20
H	6.1
O	16
N	0.5
S	0.3
Cl	< 0.1

The first reaction is the char combustion reaction, which takes place at the bottom of the gasifier where the air is injected. The gasification zone follows the combustion one. The gasification reactions are endothermic. The heat necessary to activate these reactions is in part supplied by the exothermic combustion and in part by the furnace.

The reaction rates were considered to follow an Arrhenius temperature dependence and to be proportional to the difference between the actual and the equilibrium reactant concentration:

$$r_1 = C_{RF\_C} A_1 n \exp\left(-\frac{E_1}{RT}\right)\left(P_{O_2} - \frac{P_{CO_2}^2}{K_1}\right) \quad (1)$$

$$r_2 = C_{RF\_G} A_2 n \exp\left(-\frac{E_2}{RT}\right)\left(P_{CO_2} - \frac{P_{CO}^2}{K_2}\right) \quad (2)$$

$$r_3 = C_{RF\_G} A_3 n \exp\left(-\frac{E_3}{RT}\right)\left(P_{H_2O} - \frac{P_{CO} \cdot P_{H_2}}{K_3}\right) \quad (3)$$

$$r_4 = C_{RF\_G} A_4 n \exp\left(-\frac{E_4}{RT}\right)\left(P_{H_2}^2 - \frac{P_{CH_4}}{K_4}\right) \quad (4)$$

$$r_5 = C_{RF\_G} A_5 n \exp\left(-\frac{E_5}{RT}\right)\left(P_{CH_4} P_{H_2O} - \frac{P_{CO} \cdot P_{H_2}^3}{K_5}\right) \quad (5)$$

where A and E [J.mol^{-1}] are the reaction Arrhenius parameters, n [mol m^{-3}] the molar concentration of gaseous species, C_{RF} the char reactivity factor, which was estimated from experimental data as described below and K_j the equilibrium constants for each reaction that were calculated from the Gibbs energy of reaction:

$$\ln K_j = -\frac{\Delta G_j}{RT}$$

The constants C_{RF_C} and C_{RF_G} are the char reactivity factors for the combustion and the gasification reactions, respectively. These parameters represent the reactivity of different types of char. They affect the production rate of the species and the temperature profile in the reactor. These parameters were considered different for the combustion and the gasification reactions and were estimated by fitting the temperature experimental curve as explained in the next section.

The chemical species considered in this work are CO, CO_2, CH_4, H_2, H_2O, N_2, and O_2. For each species i the mass balance equation is:

$$\frac{d(n_i\, v)}{dz} = R_i$$

where z is the bed axial coordinate, v [m.s^{-1}] the superficial gas velocity and R_i [mol.m^{-3}.s^{-1}] is the net rate of formation and is reported in table 2.

Table 2: Net rate of production of the different gaseous species

Species	R_i (mol/m^3.s)
N_2	0
O_2	$-r_1$
CO_2	$r_1 - r_2$
CO	$2\, r_2 + r_3 - r_5$
CH_4	$r_4 - r_5$
H_2O	$-r_3 - r_5$
H_2	$r_3 - 2\, r_4 + 3\, r_5$

The variation of the temperature (T [K]) on the axial coordinate was carried out using an energy balance:

$$\frac{dT}{dz} = \frac{1}{v \cdot \sum_i n_i c_i}\left(\sum_j r_j \Delta H_j - v\frac{dP}{dz} - P\frac{dv}{dz} - \sum_i R_i\, c_i\, T\right) \tag{6}$$

where $-\sum_j r_j \Delta H_j$ is the energy released by the chemical reaction j, c_i [J.mol^{-1}K^{-1}] the molar heat capacity of the species i and P [Pa] the total pressure.

Two further equations are needed to complete the reactor modelling: an equation for the pressure gradient along the bed and one for the variation of the superficial gas velocity. The pressure gradient is a function of the superficial gas velocity and it was derived from the Ergun equation [6]:

$$-\frac{dP}{dz} = \frac{150\, \mu\, (1-\varepsilon)^2}{D_P^2\, \varepsilon^3}\, v + \frac{1.75\, \rho\, (1-\varepsilon)}{D_P\, \varepsilon^3}\, v^2 \tag{7}$$

where ε is the void fraction of the bed, D_P [m] the particle diameter, μ [kg.m^{-1}s^{-1}] the fluid viscosity and ρ [kg.m^{-3}] the fluid mass density.

The expression for the velocity gradient was obtained by differentiating the ideal gas law:

$$\frac{dv}{dz} = \frac{1}{\sum_i n_i c_i + n\,R}\left(\frac{\sum_i n_i c_i \sum_i R_i}{n} - \frac{\sum_j r_j \Delta H_j}{T} - \frac{dP}{dz}\left(\frac{v}{T} + \frac{v \sum_i n_i c_i}{P}\right) - \sum_i R_i c_i\right) \qquad (8)$$

A system of ten coupled first order differential equations were obtained including ten variables: concentrations of the seven considered species, temperature, pressure and velocity.

To make the model time dependent, the variation of the bed length as a function of time was calculated experimentally, as explained in the next session.

3.1 Char reactivity factor and bed length consumption rate determination

To determine the bed length consumption rate and the CRF factor, experiments on a smaller tubular quartz reactor heated by an electrical resistance were performed. The same operative conditions, scaled to the size of the *smaller* reactor, were reproduced. Four thermocouples were disposed at 0, 0.5, 1.2 and 2.6 cm along the reactor axial coordinate. The bed consumption rate (α) was calculated as:

$$H(t) = -\alpha t + H_0 \qquad \alpha = 0.0015\ \mathrm{cm/s}$$

where $H(t)$ and $H_0 = 5$ cm are the bed length as a function of time and the bed length initial value, respectively.

A curve of the temperature profile along the reactor axial coordinate is reported in fig. 2 for three different times. The time needed for the complete bed consumption was calculated considering that the end of gasification occurs when the temperature is almost constant in the reactor. The curve at 250 s represents the beginning of the process. The combustion process occurs where the temperature is higher. Then the temperature decreases since the endothermic gasification reactions take place. At t = 2,000 s, the temperature is almost constant along the bed height and the process is considered finished.

A fitting to find the correct value of the two CRF factors was made on the temperature profile. In fig. 2, the comparison between experimental and simulated data is reported showing a good agreement. The CRF values were calculated to be 100 and 1500 for the combustion and gasification reactions, respectively.

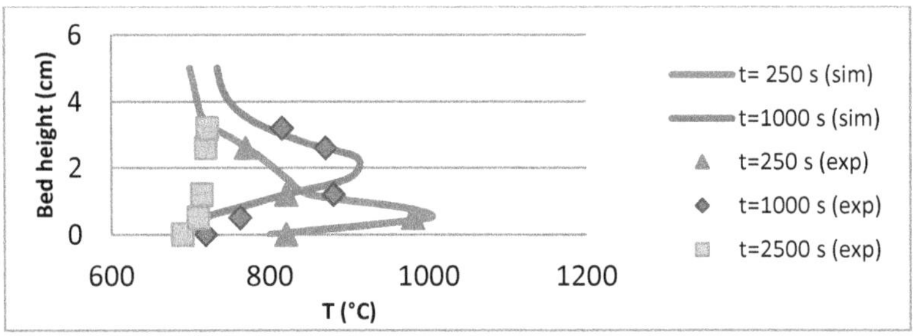

Figure 2: Temperature profile along the gasifier axis at three different residence time, experimental and simulated data.

4 Results and discussion

In an up-draft fixed-bed gasifier, the combustible and the oxidant flow counter-currently, the fuel is introduced at the top of the gasifier and the oxidant at the bottom. In fig. 3, a generic representation an up-draft gasifier with the typical temperature profile is reported.

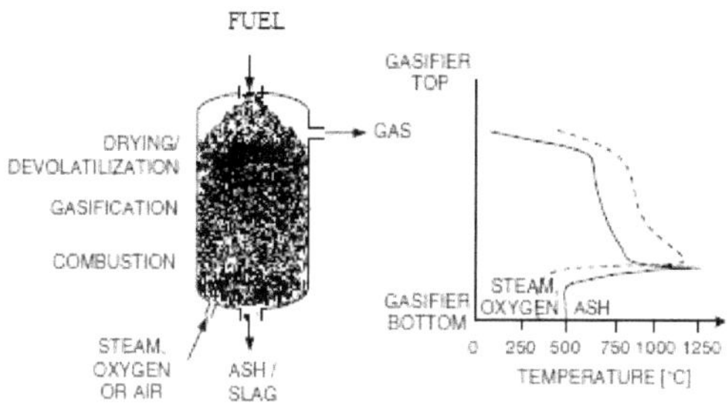

Figure 3: Up-draft gasifier and temperature profile [7].

When the fuel is injected, it is dried and then the pyrolysis process occurs. The char forms a fixed bed in the reactor where the gasification and the combustion reactions take place. The combustion zone is the first zone from the bottom of the reactor and the exothermic reactions make the process autothermal. The raw gas exits at the top of the gasifier and is laden with pyrolysis products such as tar, oil and light hydrocarbons. Thus to complete the gasification reactions and to allow the tar cracking, a reforming unit is generally placed after the gasifier.

Two types of experiments were performed. First a complete gasification, using the reactor as an updraft gasifier was carried out. In fig. 4, the concentration profiles of CH_4, CO and CO_2 of a complete updraft gasification process are reported. It can be noticed that the pyrolysis and the gasification processes occur in two separated steps: the pyrolysis process takes place in the first 1000 s, i.e. during methane production. The gasification occurs until after the CO concentration is higher than that of CO_2. The CO concentration becomes smaller than the CO_2 one as the bed height decreases and the CO_2, produced by the combustion, does not have enough time to react with the residue char.

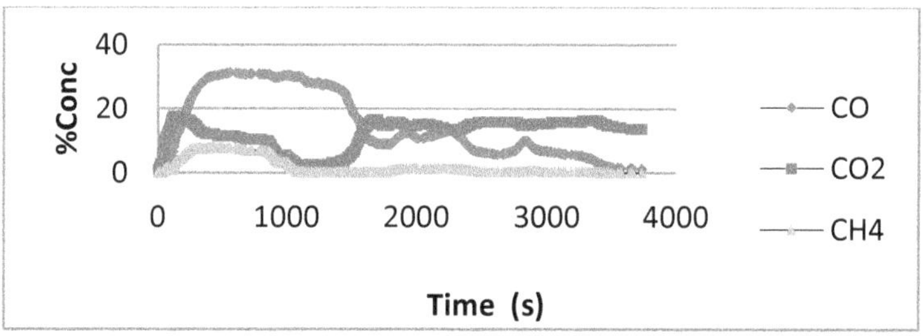

Figure 4: Concentration profile of a complete up-draft gasifier.

Since experimental data on the gasification process alone were needed, the process was split into two parts: the pyrolysis carried out at 850°C in an inert atmosphere and the gasification in a fixed bed.

The results of the pyrolysis are reported in table 3. The volatile concentrations were calculated before and after the reforming reactor. The amount of tar collected before the reforming reactor was calculated to be 30% of the volatile matter and the hydrogen concentration is 9%. The concentration of the pyrolysis gas species as a function of time after the reforming unit is reported in fig. 5.

Table 3: Pyrolysis gas and tar concentration.

	CO	CO_2	CH_4	H_2	Tar
Before reforming	21%	15%	11.4%	9%	30%
After reforming	25%	15%	16%	19%	5%

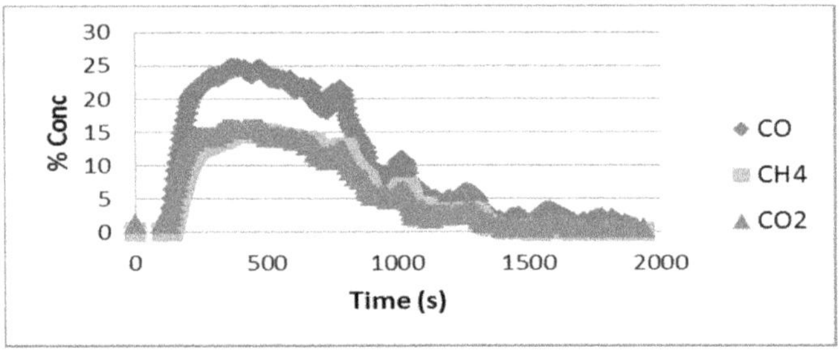

Figure 5: Concentration profile of CH_4, CO and CO_2 during the olive residue pyrolysis after the reforming reactor.

After the reforming unit, a great reduction of the tar amount that is converted in lighter species can be noticed. The reforming unit converts most of the tar enriching the gas in hydrogen and methane. Thus the reforming reactor is fundamental to reduce the tar quantity that can create problems in the following zones of the plant.

The model was validated using experimental data. In fig. 6, the comparison between simulated and experimental data is reported, showing a good agreement.

The model was used to predict the gas composition in order to optimize the operative conditions that gave the best syngas composition. In table 4, the simulated gas compositions obtained by varying the ER, the steam air ratio and the temperature are reported.

Increasing the amount of steam and the temperature and decreasing the ER ratio, the quality of the obtained syngas improves. However, using a low equivalent ratio can be a problem since the gasification process becomes too slow. As expected, increasing the steam flow, the amount of hydrogen rises. This effect could not be convenient because the gasification time increases, the reactions including the water vapour are endothermic, and the steam production is an expensive operation that can reduce consistently the plant efficiency. The best gas composition is CO 35%, CO_2 2.25% and H_2 11.5% that is a typical gas composition for air gasification.

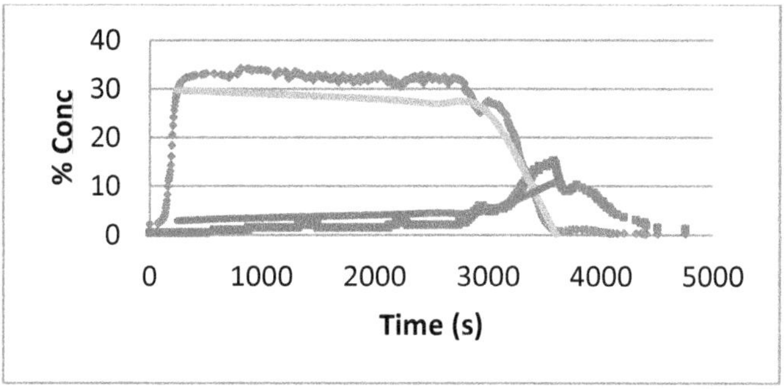

Figure 6: Experimental CO (blue line) and CO_2 (red line) profile compared with CO (green line) and CO_2 (violet line) simulated profile obtained for a ER = 0.6, T = 1100 K and 30% of steam.

Table 4: Simulated results of the syngas composition expressed in % in volume.

ER	T (K)	O_2 steam/O_2 air	%CO	%CO_2	%H_2
0.5	1000	10%	25.73	4.75	8.00
0.6	1000	10%	23.56	5.49	6.50
0.7	1000	10%	20.19	6.50	4.30
0.6	1100	10%	30.45	2.30	8.50
0.7	1100	10%	26.54	4.35	5.70
0.6	1100	20%	33.15	2.60	9.75
0.7	1100	20%	29.30	4.10	6.80
0.6	1100	30%	35.00	2.25	11.50

5 Conclusions

The gasification represents promising technology to recover energy from waste when the calorific value is low. The pyrolysis process produced a good quality gas that can be used as fuel and the reforming unit reduces considerably the tar amount that can cause problem in the following part of the plant. The syngas contains a low quantity of pollutant and can be sent to a purifying unit and then used directly as a combustible or as a raw material for other processes like the hydrogen production or the Fischer Tropsch. The composition of the obtained syngas is in line with typical syngas composition obtained by means of air gasification.

References

[1] Joung, H., Seo, Y. & Kim, K., Distribution of dioxins, furans, and dioxin-like PCBs in solid products generated by pyrolysis and melting of automobile shredder residues. *Chemosphere*, **68**, pp. 1636–1641, 2007.

[2] Eurostat 2009. Available at http://epp.eurostat.ec.europa.eu/portal/page/portal/waste/data/wastestreams/elvs. [Accessed May, 2010].

[3] Ciacci, L., Moreselli, L., Passarini, F., Santini, A. & Vassura, I., A comparison among different automotive shredder residue treatment processes. *International Journal of Life Cycle Assessment*, **15**, pp. 896–906, 2010.
[4] Barnes, G.R.G., McKibbin, R. & Giltrap, D.L., A steady state model of gas-char reactions in a downdraft biomass gasifier. *Solar Energy*, **74**, pp. 85–91, 2003.
[5] Wang, Y. & Kinoshita, C.M., Kinetic model of biomass gasification. *Solar Energy*, **51**, pp. 19–25, 1993.
[6] Ergun, S., Fluid flow through packed columns. *Chemical Engineering Progress*, **48**, pp. 89–94, 1952.
[7] Higman, C. & van der Burgt, M., *Gasification*, Elsevier, 2003.

Conversion of waste rubber as an alternative route to renewable fuel production

M. Stelmachowski & K. Słowiński
Technical University of Lodz, Poland

Abstract

Waste rubber (particularly scrap tires) generated worldwide exerts a detrimental influence on the world economy and the environment. Its fraction in municipal waste amounted to 2% by mass in 2000. By the end of 20th century, rubber production was about 34 million tons world-wide. It is estimated that 20% of tires have to be recycled every year. Dumping and land filling of used tires were the most popular methods of utilization of them until not long ago. Other alternative methods that have been used for tire recycling such as retreating, reclaiming, incineration and grinding have also significant drawbacks and/or limitations but they influence the environment to a lesser extent than dumping. Pyrolysis or gasification of waste rubber may be the best way of its utilization. The results of the thermal degradation of waste rubber performed in a new type of tubular reactor with molten metal are presented in the paper. The melting and degradation processes were carried out in an apparatus at temperatures of 390–420°C. The problems encountered with the disintegration of wastes, the heat transfer from the wall to the particles, cooking at the walls of the reactor and mixing of the molten volume of wastes are significantly reduced. Two products: gaseous (below ~ 15 weight%) and liquid (over 40 weight%) product were obtained during the degradation of waste rubber. Both streams of products were analyzed by gas chromatography. The gaseous stream contained hydrocarbons from C_2 to C_8 and the liquid product consisted of hydrocarbons C_4 to C_{24}. Over 80 mol% of liquid hydrocarbons mixture was the fraction C_4-C_{10}. The obtained liquid product may be used in petrochemical and refinery industry or may be recycled for tire manufacturing.
Keywords: thermal decomposition, waste rubber, scrap tires, molten metal.

1 Introduction

Scrap tires cause an enormous environmental and economical problem. Dumping and land filling of used tires were the worst methods of their utilization but, simultaneously, the most popular [1–5] until not a long time ago. The statistical data about the amount of waste rubber and scrap tires and the methods of their utilization are sometimes incoherent. Selected, estimated and integrated data are presented in the table 1.

Table 1: Recycling methods of scrap tires in 2004. Integrated and estimated data from different sources.

Method	UE		USA		Japan and China
	10^6Mg	%	10^6Mg	%	10^6Mg
Thermal (incineration)	0.75	30.7	1.10	43.4	Data not available
Dumping, land filling	0.95	38.9	0.87	34.3	
Civil engineering	0.28	11.5	0.29	11.4	
Direct recycling vulcanization)	0.15	6.1	0.14	5.5	
Other methods e.g. pyrolysis, cracking and chemical methods	0.31	12.7	0.14	5.3	
Total	2.44	100.0	2.54	100.0	2.2–2.6

At the end of 20th century, rubber production was about 34 million tons in the world [1–5]. A huge amount of produced rubber is used in tires production. It is estimated that 20% of tires have to be recycled every year [5]. A typical used automobile tire weighs 9–10 kg. Recoverable rubber (about 5.4–5.9 kg) consists of natural (about 35 weight%) rubber and synthetic rubber (about 65 weight%). A typical scrap truck tire weighs 18-20 kg and also contains 50–60% of recoverable rubber. The remaining part consists mainly of black carbon, zinc oxide and steel cord. These components may be also recovered.

The rubber wastes recycling industry was almost as old as the rubber production industry and over 50% of used rubber products was reclaimed in the beginning of 20th century. In 1960s and 1970s only 20% of scrap rubber was recycled and in the end of 20th century about 40–70% of used tires were dumped in massive stockpiles in many countries, which provide ideal breeding grounds for diseases and may also cause a fire hazard (e.g. in California 1999, in Poland 2003). The other alternative methods that have been used in tire recycling such as retreating, reclaiming, incineration and grinding have also significant drawbacks and/or limitations but they influence the environment to a lesser extent than dumping [1–6]. Thermal methods have been applying for rubber utilisation as well.

The thermodynamic analysis of different thermal processes points out that the best (from energy saving point of view) methods are: gasification, cracking (pyrolysis) and, finally, incineration [4]. Therefore, gasification, pyrolysis and cracking can be considered as viable recycling methods to treat the scrap tires to obtain valuable raw materials and/or energy. The scientific investigations are focused on the pyrolysis [1–3, 6, 7]. The pyrolysis gas can be used as a fuel in the pyrolysis process. A solid residue may be used for producing low-grade black carbon and the liquid product – a mixture of hydrocarbons may be used in petrochemical and refinery industry for producing fuel. Several types of reactors were invented and patented. The laboratory scale apparatus, pilot and industrial plants are generally based on tubular reactors to perform thermal pyrolysis [1–7]. Fluidized-bed reactors (for pyrolysis of disintegrated tires) have been proposed as well due to their many advantages [8–10]. However, the industrial applications of the invented reactors are rare. It means that the patented solutions are probably technically imperfect and their profitability is weak. The main weakness that should be overcome concerns the heat transfer from a heating medium to the particle of disintegrated rubber, because rubber has very low value of thermal conductivity [9, 10]. Other obstacles are connected with the utilization methods of other components of tires: black carbon, sulphur, zinc oxide and steel cord because rubber content in a tire is only about 50% by mass. Thermal conversion of rubber will be profitable only if all components of tires are recycled and the gas product of pyrolysis is desulfurized. The methods of desulfurization of gases are well known and there are also methods of utilization of black carbon [11, 12].

The problem of bad conditions of the heat transfer from a heating medium to the particle of disintegrated scrap tire may be solved by a new technology based on the thermal decomposition of waste in the molten metal bed reactor [13]. Several patents are based on this process for the degradation of organic wastes [14–17] scrap plastics. The paper presents the results of investigations on thermal rubber decomposition in molten metal.

2 Experimental work

2.1 Materials

Disintegrated waste bicycle tires and flat rubber boards were the stock for the thermal decomposition. Thirteen kilograms of alloy of tin and lead was used to create the molten metal bed. The properties of the alloy are presented in table 2.

Table 2: The composition of the alloy used in the reactor.

Fraction of tin, % weight	59–61
Fraction of lead, % weight	38–40
Fraction of impurities, % weight	1
Specific gravity, g/cm^3	8.5
Melting temperature, °C	183–185

2.2 Experimental set-up

The degradation of rubber was carried out in a new type of a tubular reactor with molten metal called "the tube in the tube". The construction of the reactor differs from the known basin reactors that have been patented until now [14–16]. It was built from two tubes of different diameters [17]. The inner (input) tube was placed coaxially within the external (outflow) tube. A mobile piston was located inside the inner pipe for transporting wastes into the molten metal bed. The hold-up time of waste rubber in volume of the liquid metal bed depends on the piston's speed of shifting. The scheme of the reactor and the scheme of the experimental set-up are presented in fig. 1.

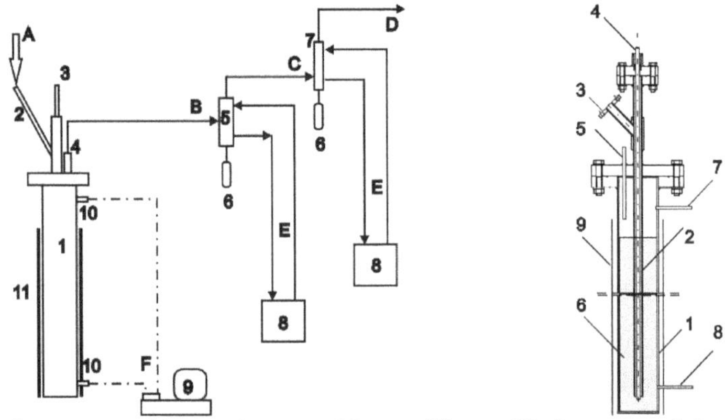

(1) the reactor with a molten metal layer, (2) the loading port, (3) the piston, (4) the outlet of vapors, (5) the basic cooler of the liquid product, (6) receivers of liquid products, (7) the final cooler, (8) thermostats, (9) the computer and acquisition system data, (10) thermocouples, (11) electrical heating. A – polymers, B, C – vapor products, D – non-condensable gases to bubble flow meter and to GC, E – cooling water, F – temperature signal to the acquisition system.

(1) the external (outflow) tube, (2) the inner (input) tube, (3) the loading port, (4) the piston for wastes transporting (5) the outlet of vapors, (6) the bed of molten metal, (7),(8) thermocouples (gas phase, molten metal), (9) electrical heating.

Figure 1: The scheme of the laboratory set-up and the vertical section of the tubular reactor.

2.3 The run description

Disintegrated rubber particles were put through the loading port to the inner tube and then slowly transported by the piston into the liquid molten alloy to the end of the inner tube. Meanwhile, rubber was melted and decomposed. Products of degradation and un-decomposed components flowed out to the external tube and next to the surface of the bed. Further degradation was carried out in the molten metal bed in the external tube and on the surface to give final products – the mixture of hydrocarbons. The vapours flowed out from the reactor were condensed in coolers. Liquid products were collected in small receivers. Gaseous product, the mixture of un-condensable hydrocarbons, flowed out from the reactor to the bubble flow meter and next through a sampling port to the ventilation system. Samples of liquid product and gaseous product were analyzed by gas chromatography. The temperatures of the molten metal bed, the gas phase and the liquid product in the coolers were measured and recorded by the acquisition data system.

3 Results and discussion

Several experiments of thermal degradation of waste rubber have been performed in the reactor with molten metal. The results of nine of them carried out in three runs are presented in the paper. The general profile and mass balances of all of them are presented in table 3. In the first run only flat rubber board was decomposed and it was performed in different way than two others. The rubber plate was cut to particles by size about (5–15) × (2–4) × 3 mm^3. Total mass of the loaded stock was 229.5 g and it was put into the reactor in three portions of 77.00, 100.01 and 120.72 g in very short intervals, one after the other. The liquid product was received in the first cooler at the temperature 408°C of the molten metal bed 8 minutes after the time of loading. The temperature in the first, basic cooler of hydrocarbons was 23°C. The liquid product was collected into several small receivers. This enabled the liquid stream product to be calculated in time. The temperature of the gaseous phase in the reactor was between 228°C and 226°C during all the experiments. The gas product was cooled in two steps. In the first cooler, heavy hydrocarbons were condensed and then the gas flowed to the second cooler, where light condensable components were received. Non-condensable hydrocarbons flowed through the sampling port and bubble flow meter to the vent system.

Table 3: The profile and mass balance of all runs.

Run number	Mass of the stock	Temperature	Mass fraction of the products		
			gas	liquid	solid (residue)
	g	°C	% mass		
1	299.5	409–440	5.79	41.04	53.17
2A	156.97 (51.43 + 30.19 + 39.33 + 36.02)	425–427	14.74	41.28	44.4
2B		391–431			
2C		365–427			
2D		379–430			
3A	153.37 (38.02 + 38.38 + 38.39 + 38.58)	400–403	14.73	43.62	41.64
3B		403			
3C		396–403			
3D		395–403			

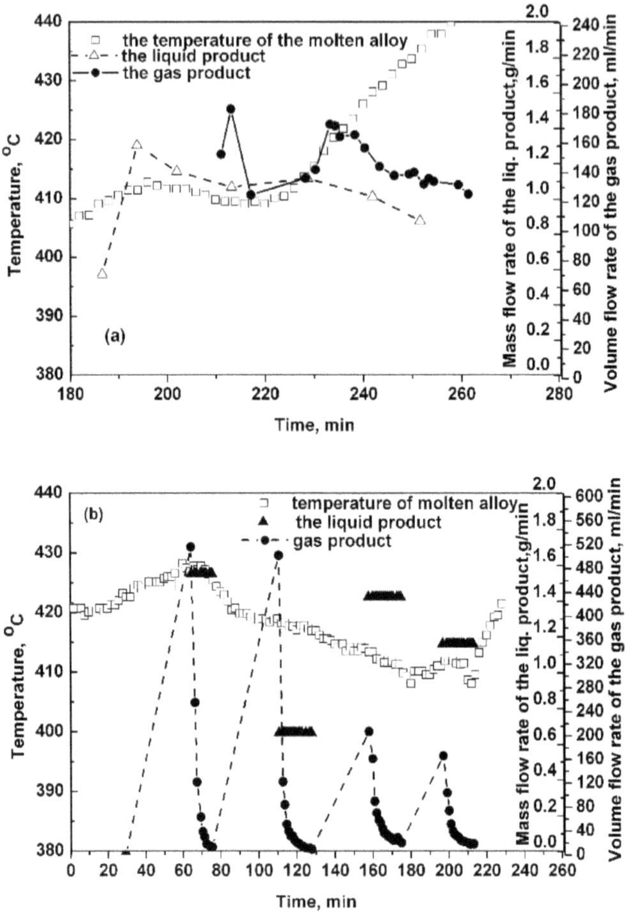

Figure 2: The distribution of the temperature of the molten alloy in the reactor and flow rates of the gas and liquid products in the runs 1 and 3.

Runs 2 and 3 consisted of four experiments each. The stock consisted of scrap bicycle's tires. They were cut to particles of size between 5 and 10 mm. The total stock was divided in four portions, in both runs. When the liquid metal obtained demanded temperature, the first small part of disintegrated rubber particles was input through a loading port into the reactor and then transported into molten metal by the piston. The gaseous product was obtained in a few minutes after loading and in further few minutes the liquid product was collected to only one collector for each experiment (each decomposed portion of rubber). When the flow of product was decreased, the temperature of the molten metal bed was changed. When new conditions were achieved another part of rubber waste was loaded to the reactor for the next experiment.

The profile of product flow rates, the temperature distribution of molten metal for the run 1 and 3 is presented in fig. 2, as a representative example.

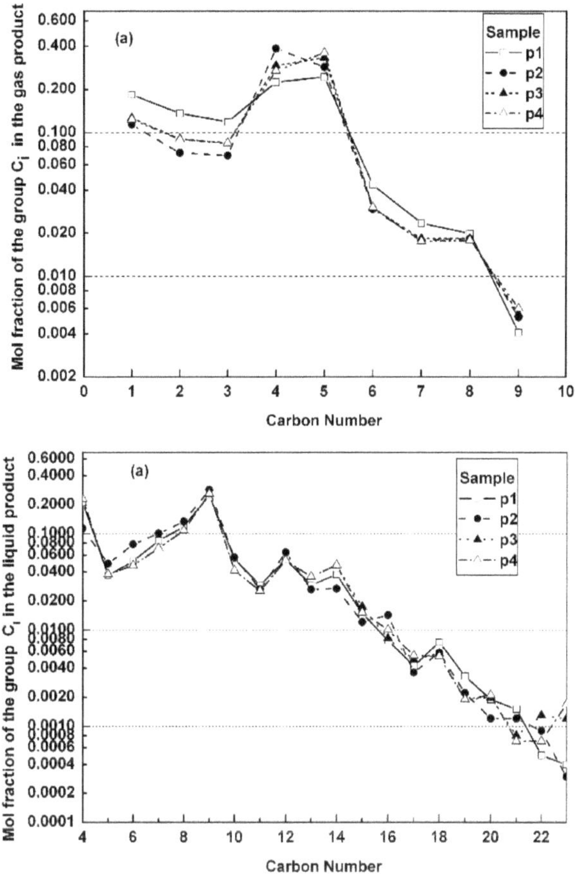

Figure 3: The carbon number gas product (a) and liquid product distribution for rubber decomposition in the run 3.

Runs 2 and 3 were very similar. The method of performing the runs and the amount of the stock material were almost the same. The liquid and gas product distributions for run 4 are presented in fig. 3. It is also representative of all experiments. The compositions of the gas and liquid samples were similar and stable during all runs as it can be seen in fig. 3. The lines present the GC analysis results of different samples collected with the time. Run 1 was carried out in another way and under other conditions than those of runs 2 and 3 but the average liquid products obtained from all experiments are very similar and stable for all runs as shown in fig. 4. The fraction compositions of average liquid samples collected in all runs are presented in table 4. The liquid products obtained from thermal decomposition of waste rubber in the molten metal contained above 95% mol most valuable, light fractions of hydrocarbons (called "gasoline" (C_4-C_{10}) and "diesel" (C_{11}-C_{16}) fraction. The composition of the liquid product makes it usable as a stock in petrochemical or rubber industry.

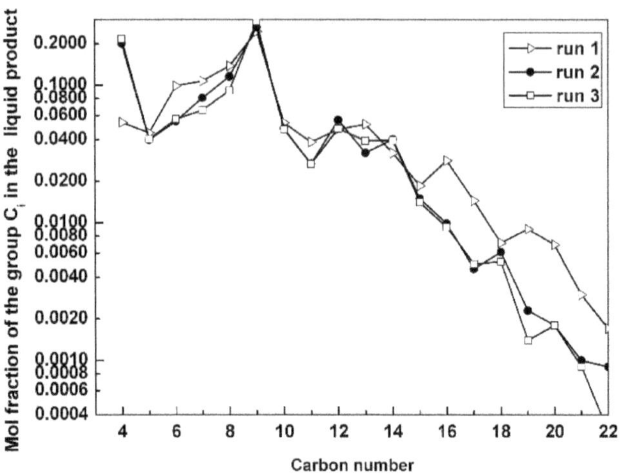

Figure 4: The carbon number average liquid product distribution for rubber decomposition in the runs: 1, 2 and 3.

Table 4: Fraction composition of the average liquid products.

Run number		1	2 (A-D)	3 (A-D)
Fraction		Fraction composition of the liquid product		
C_4-C_{10}	% mol	68.8	75.3-75.8	73.6-77.7
C_{11}-C_{16}	% mol	24.2	21.2-22.0	19.6-24.2
C_{17}-C_{24}	%mol	7.0	2.6-2.9	2.0-3.0
paraffin	% mol	31.5	27.5-32.3	38.5-39.7
olefins	%mol	68.5	65.7-72.6	60.3-62.5

The received products are generally the mixture of paraffin and olefins. The ratio of paraffin to olefins is presented in table 4 too for liquid products. The amount of olefins is about 50% greater than that of paraffin. The liquid products are fluids in ambient conditions; however, black carbon is suspended in a liquid mixture. Therefore, before further use, it has to be clarified.

The qualitative analysis of the liquid samples made by FTIR proved that the content of aromatic hydrocarbons was below 5 mol.%.

The content of heavy metals in liquid product was analyzed too and it was made by AAS. The concentration (in liquid) of lead (Pb) and tin (Sn) was between 20 and 30 mg/kg, of zinc below 17 mg/kg, of nickel (Ni) and chromium (Cr) below 1.5 mg/kg. The relative analytical errors of metals determination were: 0.8–1.0% (for Pb, Sn, Zn) and 4% (for Ni and Cr).

4 Conclusions

The thermal decomposition of rubber wastes performed in a reactor with molten metal may have many advantages. The process is fast. The problems with the heat transfer resistance between the particles of wastes and the heating medium is overcome due to direct contact between rubber and molten metal. Therefore, the cooking processes may be minimized. The ratio of the amount of gaseous product to the amount of liquid product is less than that resulting from other thermal decomposition processes. The liquid product contains mainly gasoline fraction of hydrocarbons. The hold-up time of wastes in the reactor and the reaction time are shorter than in classical reactors applied for rubber pyrolysis. Therefore the amount of the recombination reactions products is minimized and the fraction of aromatic hydrocarbons is smaller than in other processes based on pyrolysis in tubular or fluidized rectors. The content of heavy metals is small that means that the process temperature is proper and the metal particles and metal ions have not been transferred to the liquid products. However, thermal cracking processes of rubber in the reactor with molten metal bed (as other thermal processes) will be cost-effective if all of the obtained products: not only hydrocarbons but, for example, black carbon as well may be recycled and used for producing fuel and other useful semi-products. About 10% of recovered energy has to be used (recycled) for the disintegration of waste rubber.

References

[1] Galvagno, S., Casu, S., Casablanca, A. & Cornacchia, G., Pyrolysis for treatment of scrap tyres: preliminary experimental results. *Waste Management*, **22(8)**, pp. 917-923, 2002.

[2] Rodriguez del Marco, I., Laresgoiti, M.F., Cabrero, M.A., Torres, A., Chomon, N.J. & Caballero B., Pyrolysis of scrap tires. *Fuel process Technology*, **72**, pp. 9-22, 2001.

[3] Williams, P.T. & Brindle, A.J., Catalytic pyrolysis of tyres: influence of catalyst temperature. *Fuel*, **81**, pp. 2425-2434, 2002.

[4] Stelmachowski, M., Termo-katalityczna degradacja polimerów, (Thermo-catalytic degradation of polymers – monograph in polish), Monografie, PAN Oddział w Łodzi, Komisja Ochrony Środowiska, Łódź, 2003, ISBN – 83-86492-19-8.

[5] Sharma, V.K., Fortuna, F., Mincarini, M., Berillo, M. & Conacchia, G., Disposal of waste tires for energy recovery and safe environment. *Applied Energy,* **65**, pp. 381-394, 2000.

[6] Lin, I.-P., Chang, Ch.-Y., Wu, Ch.-H. & Shih, S.-M., Thermal degradation kinetics of polybutadiene rubber. *Polymer Degradation and Stability*, **53,** pp. 295-300, 1996.

[7] Darmstadt, H., Roy, Ch. & Kallaguine, S., Characterization of pyrolytic carbon blacks from commercial tire pyrolysis plants. *Carbon*, **33(10)**, pp. 1449-1455, 1995.

[8] Dai, X., Yin, X., Wu, Ch., Zhang, W. & Chen, Y., Pyrolysis of waste tires in a circulating fluidized-bed reactor. *Energy*, **26,** pp. 385-399, 2001.
[9] Kaminski, W., The Hamburg fluidized-bed pyrolysis process to recycle polymer wastes and tires. *Feedstock Recycling and Pyrolysis of Waste Plastics: converting Waste Plastics into Diesel and other Fuels*, eds. J. Scheirs & W. Kaminski, John Wiley & Sons, Ch. 17, pp. 475-490, 2006.
[10] Yang, J., Tanguy, P.A. & Roy, C., Heat transfer, mass transfer and kinetics study of the vacuum pyrolysis of a large used tire particle. *Chemical Engineering Science,* **50(11)**, pp. 1909-1922, 1995.
[11] Ko, D.C.K, Mui, E.L.K., Lau, K.S.T. & McKay, G., Production of activated carbons from waste tire – process design and economical analysis. *Waste Management,* **24,** pp. 875-888, 2000.
[12] Benallal, B., Roy, C., Pakdel, H., Chabot, S. & Poirier, M.A., Characterization of pyrolytic light naphta from vacuum pyrolysis of used tyres. Comparison with petroleum naphtha. *Fuel,* **74(11)**, pp. 1589-1594, 1995.
[13] Newborough, M., Highgate, P. & Vaughan, P., Thermal depolymerisation of scrap polymers. *Applied Thermal Engineering,* **22(17)**, pp. 1875-1883, 2002.
[14] Domingo, J. & Cabanero, D., Process and device for regeneration of monomer from polymethyl methacylate, Spanish patent 192909, 1949.
[15] Mausre, W.E., Donahue, J.R., Larue, G.W., Bonney, L.H., Glanton, G.W. & Harris, W.L., Method and apparatus for thermal conversion of organic matter. US 342056 and European Patent Application 90401093.1, 1989.
[16] Stelmachowski, M. & Tokarz, Z., The method of continuous conversion of waste plastic, Patent application (polish) P358773, 2003.
[17] Stelmachowski, M., The reactor and method for thermal conversion of waste polymers, Patent application (polish) P384806, 2008.

Process simulation and sensitivity analysis of waste plastics gasification in a fluidized bed reactor

P. Kannan, A. Al Shoaibi & C. Srinivasakannan
The Petroleum Institute, United Arab Emirates

Abstract

This paper presents a *simplified* process simulation model of typical waste polyethylene gasification in a fluidized bed reactor using Aspen Plus. The proposed model incorporates both physical and chemical processes, including drying, pyrolysis, combustion and gasification, using various *inbuilt* modules to predict the resulting product gas composition and temperature. A detailed sensitivity analysis investigating the effects of various process parameters, including equivalence ratio, bed height and steam/fuel ratio has been presented.
Keywords: gasification, waste LDPE, Aspen Plus, modelling and simulation.

1 Introduction

Gasification in commercial scale is practiced based on batch, semi batch and continuous modes of operation depending upon the processing capacity of the plant. Typically a plant processing large throughput utilize fluidized beds due to the advantages such as enhanced gas-solid contact, excellent mixing characteristics [1], operating flexibility [2] and ease of solids handling [3] that lead to a better overall gasification efficiency. Although there are many different types of fluidized beds available for gasification and combustion, the bubbling fluidized type is the most preferred type whenever steam is used as a gasifying medium [4].

In general, the process of gasification for energy extraction from solid carbon source involve three simultaneous or competing reactions namely combustion, pyrolysis and gasification. The partial combustion of solid fuel creates an oxygen devoid, high temperature condition within the reactor which promotes the pyrolysis reaction, breaking the fuel into products that are a mixture of char and volatiles containing small and long chain hydrocarbons. The presence of gasifying agent (air and steam) drives the water shift reaction converting the carbon sources into a mixture of valuable chemicals, tar, fuel gases and some residual particulate matter.

In spite of numerous studies relating to extracting energy from waste, it still remains inconclusive how the feedstock composition and the fluid bed process conditions play a role in the resulting product composition and yield. In order to exert better reliability of the system, the operating variables have to be optimized and controlled with significant accuracy. The cheapest and most effective technique to qualitatively understand the effect of each operating variable and to identify possible optimal conditions is through process simulation. Such attempts at developing simulation models for process optimization has been reported in open literature of fuel sources such as tyre [3], coal [5–8], and biomass [4, 9–11] using various computer simulation packages. However, the utility of any process simulation tool has not been well explored or recorded in the literature for simulating plastics gasification in a fluidized bed.

The primary goal of the present work is to successfully test and demonstrate the applicability of Aspen Plus to the simulation of the gasification process for one of the most abundantly used plastic, polyethylene (PE). The simulation aims to provide preliminary qualitative and quantitative information on the overall behavior of the gasification process including the sensitivity to process parameters. More importantly, this model was developed with the aim of supporting the design phase of the pilot plant gasifier.

2 Model development

2.1 Assumptions

The following assumptions were adopted in the gasification model development:

- Steady state process;
- Char from pyrolysis contains only carbon and ash;
- Major components of the volatiles are hydrogen, CO, CO_2, methane, and water;
- Gasification process initiates in the bed and completes in the freeboard.

2.2 Reaction chemistry

The gasification of polyethylene in the fluidized bed involves a series of physical and chemical processes, including drying, pyrolysis or devolatilization, oxidation, and finally char reduction. As soon as the material enters the gasifier, it is rapidly dried to remove the bounded moisture. The complex material then starts to devolatilize and yields volatile matter and residual amounts of char. The char present in the gas phase is reformed into a combustible gas mixture viz. H_2, CO, CO_2, CH_4, small hydrocarbons via reactions with steam, air, and carbon-dioxide according to the environment and kinetic mechanism. While the main reduction reactions are assumed to be initiated in the reactive zone (bed), other processes continue to occur at the freeboard region of the gasifier.

The chemical reactions depicting the char oxidation and gasification along with the kinetic parameters are shown in table 1. The most widely referenced kinetic data have been considered in this work.

Table 1: Gasification reactions and their kinetic parameters (*K* has appropriate units depending on the order of the reaction and *E* is in kJ/kmol).

No.	Reaction	Rate equation	Ref
1	$C + H_2O \rightarrow CO + H_2$	$2.0 \times 10^5 \exp(-6000/T)$	[12]
2	$C + 2H_2 \rightarrow CH_4$	$0.12 \exp(-17921/T)$	[12]
3	$CH_4 + H_2O \rightarrow CO + 3H_2$	$6.1 \times 10^{14} \exp(257000/RT)\ C_{CH4}.C_{H2O}$	[5]
4	$CO + H_2O \rightarrow CO_2 + H_2$	$10^6 \exp(-6370/T)$	[12]
5	$CO + 3H_2 \rightarrow CH_4 + H_2O$	$312 \exp(30000/RT)\ C_{CO}.\ C_{H2}^3$	[5]
6	$C + O_2 \rightarrow CO_2$	$5.7 \times 10^9 \exp(38200/RT)\ C_{O2}^{0.78}$	[13]
7	$C + .5\ O_2 \rightarrow CO$	$5.7 \times 10^{11} \exp(55000/RT)\ C_{O2}^{0.78}$	[13]
8	$C + CO_2 \rightarrow 2\ CO$	$1.6 \times 10^{12} \exp(186308/RT)$	[14]
9	$CO_2 + H_2 \rightarrow CO + H_2O$	$1923 \exp(113070/RT)$	[12]
10	$2CO + O_2 \rightarrow 2CO_2$	$4\times10^{11}\exp(167269/RT)C_{CO}.C_{O2}^{0.25}.C_{H2O}^{0.5}$	[15]
11	$CH_4 + 2O_2 \rightarrow CO_2 +$	$1.6 \times 10^{10} \exp(202387/RT)\ C_{CH4}^{0.7}.C_{O2}^{0.8}$	[15]
12	$2H_2 + O_2 \rightarrow 2H_2O$	$2.2 \times 10^9 \exp(109137/RT)\ C_{H2}.\ C_{O2}$	[15]

2.3 Aspen Plus model

The *simplified* process model developed in this study encompasses only the major physical (drying and separation) and chemical processes (decomposition, combustion and reduction) involved in the gasification process. Other downstream processes including syngas purification, gas concentration and waste disposal or recirculation systems are not included in the scope of the model.

A schematic of the fluidized bed gasification process simulation developed in this study is illustrated in fig. 1. The major initial conditions of the various components of the Aspen Plus modules used in the development of the gasification model are summarized in table 2.

2.4 PE pyrolysis

A yield based Aspen Plus reactor, RYIELD, was used to simulate the decomposition of the dried polyethylene sample. The complex pyrolysis of PE is modeled as a chemical reaction with a product stream solely consisting of PE constituent components, namely carbon, hydrogen, sulphur, nitrogen. In this module, the product yield distribution as defined by the ultimate analysis shown in table 3 is specified along with the process conditions of temperature, pressure, and/or heat duty.

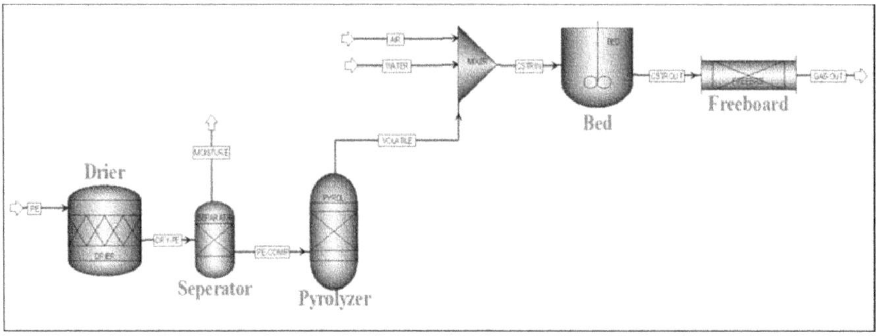

Figure 1: Aspen Plus flowsheet of the fluidized bed gasification process.

Table 2: Summary of initial values of major process parameters of various Aspen Plus modules utilized for gasification model development.

Subprocess	Aspen module	Input parameters
Drying	RSTOIC	• Temperature = 110°C • Pressure = 1 atm • Product composition according to proximate analysis
Separation	Sep	Split fraction of dried PE in outlet stream
Pyrolysis	RYIELD	• Base case Temperature = 800°C • Pressure = 1 atm • Product composition according to Ultimate analysis
Gasification	RCSTR	• Heat duty = 0 cal/s • Pressure = 1 atm • Base case Reactor volume = 0.008 m^3 • Complete reaction set
	RPFR	• Adiabatic reactor • Pressure = 1 atm • Reactor diameter = 150 mm • Reactor length = 450 mm • Complete reaction set

2.1 PE gasification

Char and volatiles resulting from pyrolysis are subjected to gasification in the bed and in the freeboard section of the gasifier. The material stream exiting from RYIELD is mixed with air and steam and transferred to the gasifying section. Using Aspen's inbuilt reactor modules RCSTR the bubbling fluidization regime in the bed was modeled as a well-mixed stirred tank reactor while the freeboard was represented as RPFR, as a plug flow reactor. In both these modules, char gasification and volatile reactions was described by reaction kinetics, as mentioned previously in this article.

Table 3: Polyethylene sample characteristics used in this study (values taken from He *et al.* [16]).

Sample	Proximate analysis (wt %)				Ultimate analysis (wt %)				
	Moisture	FC	Volatiles	Ash	C	H	N	Cl	S
Waste PE	0.02	0	99.85	0.15	85.81	13.86	0.12	0	0.06

3 Results and discussion

3.1 Effect of equivalence ratio

The equivalence ratio (ER) is a measure of the amount of air used for gasification in relation to complete combustion. It is defined as the ratio of air/fuel ratio supplied to the stoichiometric air/fuel ratio needed for complete combustion of carbon and hydrogen present in the fuel. Altering the equivalence ratio results in significant changes in the gas phase temperature and in turn gasification efficiency. ER was varied between 0.05 and 0.25 by manipulating the air flow to the process at a fixed steam and fuel flow rate. Fig. 2 illustrates the effect of ER on the product composition, exit gas temperature, and gasification efficiency of LDPE pyrolyzed at 700°C. As expected, an increase in ER increases product gas temperature due to the exothermic nature of combustion reactions. The product molar composition of hydrogen is a combined representation of the hydrogen actually formed in the gasification process, unreacted hydrogen from the fuel and that dissociated from the process steam. It can be noticed that CO efficiency attains a maximum at an ER of 0.25 corresponding to an air flow rate of 35 kg/h. However, no significant change could be observed in hydrogen efficiency with increase in ER. This indicates that the drop in hydrogen composition is due to the corresponding increase in CO production due to Reaction (7) rather than he gasification reactions which otherwise would have produced significant proportion of hydrogen. Also, the composition of syngas ($CO + H_2$) in the product stream remains unchanged beyond an ER of 0.25. Hence, the optimum ER for LDPE gasification using fluidized bed reactor is around 0.25, beyond which the process is not viable based on both technical and economic considerations.

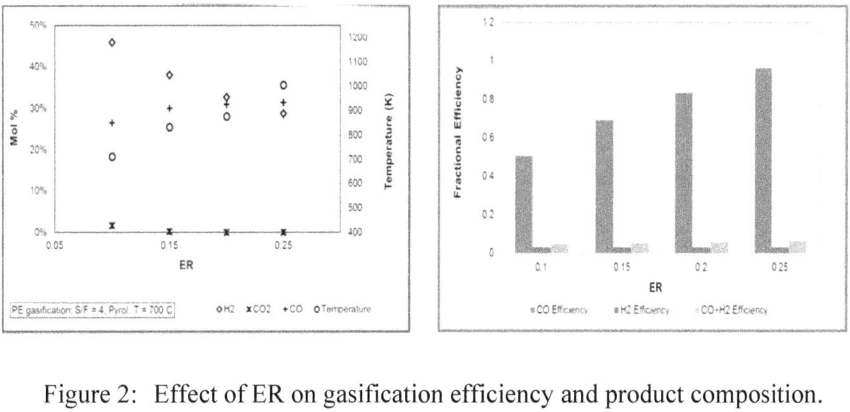

Figure 2: Effect of ER on gasification efficiency and product composition.

3.2 Effect of steam-to-fuel ratio

The primary role of steam in the gasification reactions is not only to enhance the yield of syngas, but also to control the selectivity of hydrogen and carbon monoxide. The effect of steam-to-PE ratio on the gasification efficiency was assessed in the range of 0.1 to 0.8 by altering the steam flow rate to the gasifier at a fixed air and fuel flow rate. Fig. 3 illustrates the effect of steam on the product composition and the temperature of the product stream at a bed temperature of 700°C. It can be noticed that there is a steady drop in temperature with increasing steam content undoubtedly due to the increased participation of the endothermic gasification reactions. Both the compositional trend of CO and H_2 shows an increase until the steam-to-PE ratio is 0.5, beyond which shows marginal decrease. The identical nature of the compositional curves of CO and H_2 is further evidence to the domination of water gas reactions involving carbon, CO, CH_4 and water. Also, with increasing steam flow, higher CO efficiency is attained with a significant decrease in hydrogen efficiency. It should be noted that the reason for decrease in hydrogen efficiency (in spite of *nearly* constant hydrogen product composition) is related directly to the increased steam flow and by definition of hydrogen efficiency as per eqn (3) (table 1). It can be concluded that an optimum range of 0.3 to 0.5 is desired, not only for maximizing the selectivity of CO and hydrogen, but also to have a product stream with good thermal energy suitable for practical purposes.

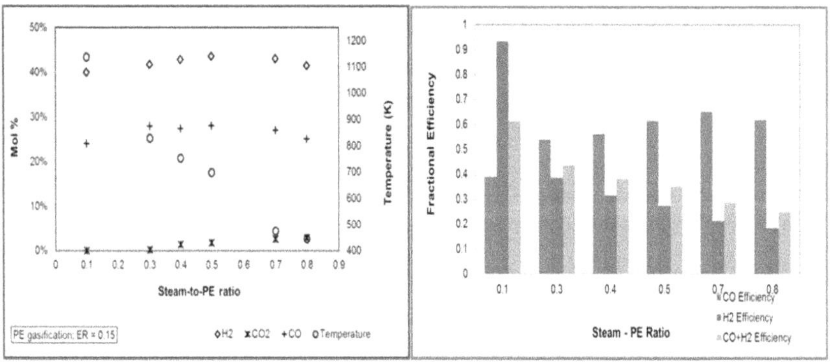

Figure 3: Effect of steam-to-fuel ratio on gasification efficiency and product composition.

3.3 Effect of bed height

The effect of bed height on gasification efficiency was investigated in terms of the ratio of bed volume to total gasifier volume wherein a value of zero corresponds to a condition of the entire gasifier considered as a freeboard (PFR) and a value of 1 would correspond to a complete bubbling bed (CSTR). The focus of this parametric analysis is to assist in design evaluation of a pilot plant gasifier. To simulate this parameter for values between 0 and 1, the bed volume (RCSTR) and freeboard volume (RPFR) were varied keeping the total volume of the gasifer as constant. For the case of bed/total volume ratio of 0, the two reactors were replaced by a single PFR equivalent to total gasifier volume whereas for case of 1, a RCSTR was considered with the appropriate volume. Fig. 4 show the effect of bed/total volume ratio on the composition and gasification efficiency. It can be noticed that there is no significant change in product molar composition although the CO efficiency seem to remain highest at a complete PFR condition. Fig. 5 depicts the effect of bed/total volume ratio on the feedstock conversion which was calculated from the amount of fuel fed and amount unreacted. It can be noticed that the conversion decreases with increase in bed volume or bed height since a PFR would yield better conversion than a CSTR. In reality, typically in a fluidized bed the conversion of plastic is envisaged only in the bubbling bed (CSTR) as the free board will only have the pyrolysed volatile particles, which undergo further the gas phase conversion. The fluidization velocities will be far less to facilitate a plug flow mode of the PE through the free board. A PFR mode of conversion in the fluidized bed is possible only at fluidization gas velocities in excess of the single particle transport velocity, which can be achieved only if the feedstock to the

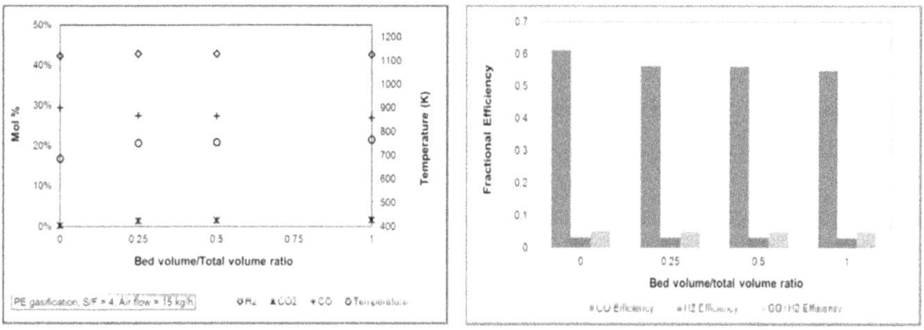

Figure 4: Effect of bed volume on gasification efficiency and product composition.

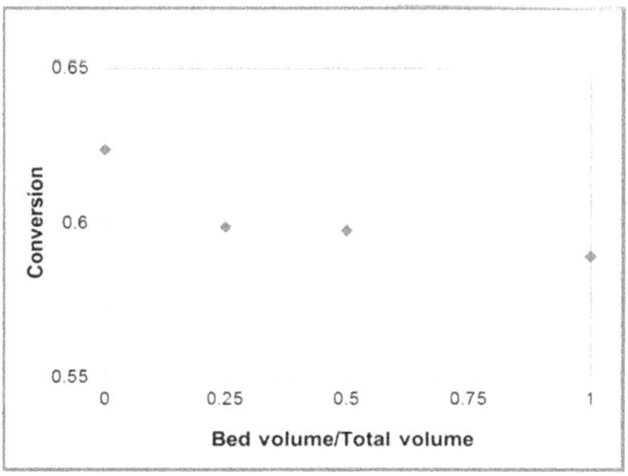

Figure 5: Effect of bed volume on PE conversion efficiency.

fluidized beds is fine powder. On the other hand operating the total volume as CSTR is not possible as distributor pressure drop should increase proportional to the bed pressure drop in order to ensure proper fluidization. In general a pressure drop of 15–20% of the bed pressure drop should be accounted for the distributor plat pressure drop in order to maintain proper fluidization.

4 Conclusions

An attempt was successfully made to simulate the plastics gasification process with a *simplified* Aspen Plus model that incorporates the significant chemical processes, namely pyrolysis, combustion and gasification using inbuilt process modules. In the course of model development, several process parameters, including equivalence ratio, steam-to-fuel ratio, and bed height were subjected to a sensitivity analysis. The following conclusions were deduced from the sensitivity analysis:

- Equivalence ratio significantly affects the product gas temperature and, in turn, CO efficiency. An optimum value of about 0.25 is required to achieve a product gas that is rich in syngas and significant thermal energy.
- Increasing the steam content to a certain extent increases the product composition of syngas; however beyond 0.5 the selectivity drops with a significant loss in thermal energy.
- Although bed height does not have a considerable effect on the product composition or gasification efficiency, it affects feedstock conversion. With increasing bed height, the conversion was found to decrease and an optimum value in the range of 0.4 to 0.5 should be considered in order to attain reasonable conversion.

Acknowledgement

The authors would like to extend appreciation to Borouge Plastics for supporting this project.

References

[1] Sadaka, S.S., Ghaly, A.E. & Sabbah, M.A., Two phase biomass air-steam gasification model for fluidized bed reactors: Part I – model development. *Biomass and Bioenergy*, **22(6)**, pp. 439-462, 2002.

[2] Arena, U., Zaccariello, L. & Mastellone, M.L., Tar removal during the fluidized bed gasification of plastic waste. *Waste Management*, **29**, pp. 783-791, 2009.

[3] Mitta, N.R., Ferrer-Nadal, S., Lazovic, A.M., Parales, J.F., Velo, E. & Puigjaner, L., Modelling and simulation of a tyre gasification plant for synthesis gas production. *Computer Aided Chemical Engineering*, **21**, pp. 1771-1776, 2006.

[4] Shen, L., Gao, Y. & Xiao, J., Simulation of hydrogen production from biomass gasification in interconnected fluidized beds. *Biomass & Bioenergy*, **32(2)**, pp. 120-127, 2008.

[5] Robinson, P.J. & Luyben, W.L., Simple dynamic gasifier model that runs in Aspen Dynamics. *Industrial & Engineering Chemistry Research*, **47(20)**, pp. 7784-7792, 2008.

[6] Yan, H.-M. & Rudolph, V., Modeling a compartmented fluidized bed coal gasifier process using ASPEN PLUS. *Chemical Engineering Communications,* **183**, pp. 1-38, 2000.

[7] Lee, H.G., Kim, C., Han, S.H. & Kim, H.T., Coal gasification simulation using Aspen Plus. *US - Korea Joint Workshop on Coal Gasification Technology*, 1992.

[8] Douglas, P.L. & Young, B.E., Modelling and simulation of an AFBC steam heating plant using ASPEN/SP. *Fuel*, **70(2)**, pp. 145-154, 1991.

[9] Mansaray, K.G., Al-Taweel, A.M., Ghaly, A.E., Hamdullahpur, F. & Ugursal, V.I., Mathematical modeling of a fluidized bed rice husk gasifier: part I - model development. *Energy Sources*, **22(1)**, pp. 83-98, 2000.

[10] Nikoo, M.B. & Mahinpey, N., Simulation of biomass gasification in fluidized bed reactor using ASPEN PLUS. *Biomass and Bioenergy,* **32(12)**, pp. 1245-1254, 2008.

[11] Doherty, W., Reynolds, A. & Kennedy, D., The effect of air preheating in a biomass CFB gasifier using ASPEN Plus simulation. *Biomass & Bioenergy*, **33(9)**, pp. 1158-1167, 2009.

[12] Inayat, A., Ahmad, M.M., Mutalib, M.I. & Yunus, M.K., Kinetic modelling of biomass steam gasification system for hydrogen production with CO_2 adsorption. *Proceedings of International conference for Technical Postgraduates*, Kuala Lumpur (Malaysia), 2009.

[13] Cuoci, A., Faravelli, T., Frassoldati, A., Grana, R., Pierucci, S., Ranzi, E. & Sommariva, S., Mathematical modelling of gasification and combustion of solid fuels and wastes. *Chemical Engineering Transactions*, **18**, pp. 989-994, 2009.

[14] Ray, R., & Thorpe, R.B. A comparison of gasification with pyrolysis for the recycling of plastic containing wastes. *International Journal of Chemical Reactor Engineering*, **5(1)**, Doi: 10.2202/1542-6580.1504, 2007.

[15] Wurzenberger, J.C., Wallner, S. & Raupenstrauch, H., Thermal conversion of biomass: comprehensive reactor and particle modelling. *AIChE Journal*, **48(10)**, pp. 2398-2411, 2002.

[16] He, M., Xiao, B., Hu, Z., Liu, S., Guo, X., & Luo, S., Syngas production from catalytic gasification of waste polyethylene: Influence of temperature on gas yield and composition. *International Journal of Hydrogen Energy*, **34**, pp. 1342-1348, 2009.

Recovering energy from liquid sanitary waste through direct alcohol fuel cells

V. Pelillo & D. Laforgia
University of Salento, Italy

Abstract

The evolution of energy systems can take place through different ways and, sometimes, by using processes already known but needing to be improved by continuous research and development. This is the case of direct alcohol fuel cell (DAFC), known since a century ago, but whose operating mode consists of employing (under specified conditions) alcohol. The objective of this work is not only to show the sustainability of these devices but mainly their application for supplying re-cycled ethanol and glycol, recovered from sanitary liquid wastes, to be used for biomedical apparatuses and plant located inside hospital pavilions, such as neonatology, resuscitation and surgery rooms, where, electric energy must not be interruptible for safety reasons and, as a consequence, DAFC must always be working. The proposed solution is supported by the U.E. Act (Directive 2006/12/CE) that allows waste recovery in order to produce energy. It is an innovating and alternative manner to provide a good use of sanitary liquid wastes instead of sending them to common and regular disposal.
Keywords: sanitary wastes, cell fuel, disposal, waste recovery, waste legislation.

1 Introduction

Hospital or clinical or sanitary wastes require particular management procedures different from those applied to general wastes. There are specific classifications according to each western country's legislation. They include not only the so-called conventional wastes but also the so-called unconventional ones, that is, radioactive wastes. To dispose these wastes, the main technique consists of using appropriate incineration plants capable of achieving the task in a way such that the environment is protected. Various studies suggest that as much as 50% of waste sent for incineration as clinical waste is in fact general waste, leading to unnecessarily high disposal costs [1]. Improved segregation can generate substantial savings. In an ideal world, segregation of waste should take place at the point of production, for example, on the wards or in operating rooms. Encouraging segregation at production sites may help to save money and provide a secure and correct way of disposing hazardous materials in order to prevent irreversible damage to the environment. For the purposes of this research, the acceptation sanitary waste is used because it includes any waste produced by any establishment where health activities are operated for the interest of humans and animals. However, the meaning of acceptations hospital and clinical waste could be a little bit confusing. Firstly, it is helpful to recall the origin of sanitary wastes and their contents. Sanitary wastes come from various sources such as private and public hospitals, health centers, research facilities, health laboratories, veterinary surgeries and dental surgeries [2]. Moreover, the contents consist wholly and partly of various items such as syringes, needles, glasses, human or animal tissue, swabs, dressings, excretions, blood, physiological fluids, pharmaceutical wastes and incontinence bags [3]. Sanitary liquids play a key role in a hospital, especially if they contain solvents that could be extracted in order to be regenerated for recycling in a new process. Improving the mechanism of waste segregation means money savings and keeping the environment safe. Different western countries, before the introduction of new regulations, namely the European Waste Code (EWC) in its diverse versions, have faced in many ways sanitary waste issues. The increasing costs and technology development may impose a new vision and a new approach to the search for ways to segregate waste in the hospital facilities. This segregation must be enhanced especially by separating liquids from solids, hazardous from non-hazardous, flammable from inflammable. The UK [4], France [5] and Italy [6], for example, regulated, in an appropriate way, the matter in the 90s by distinguishing between the types of sanitary waste as well as the procedures for final disposal according to WHO [7].

There has been an underlying disparity across Europe in the approach to the management of sanitary wastes, both within hospitals [8] and in the commercial sector. Differences have been harmonized by the European Hazardous Waste Directive 91/689/EC (HWD), which seeks to provide a precise and uniform European-wide definition of hazardous wastes, including clinical wastes, and to ensure its regulation and correct management. The classification of a waste as hazardous has considerable impact in determining how that waste is regulated, and thereby on the care required at all stages from initial disposal to final destruction. A detailed classification of wastes is set out in the European Waste Catalogue 2000/532/EC (EWC), an essential first step in the implementation of the HWD. EWC classification permits down-regulation of some clinical wastes. The EWC is pivotal to this legislation, enabling classification of wastes based upon composition and hazard (table 1). Non-hazardous wastes escape stringent control in disposal that had previously been applied in blanket fashion, permitting correspondingly lower disposal costs.

Table 1: European Waste Catalogue (2002) categorization of some sanitary wastes.

18	Wastes from human and animal health care and/or related research (except kitchen and restaurant wastes not arising from immediate health care)	
18 01	Wastes from natal care, diagnosis, treatment or prevention of disease in humans	
18 01 01	Sharps (except 18 01 03)	
18 01 02	Body parts and organs including blood bags and blood preserves (except 18 01 03)	
18 01 03*	Wastes whose collection and disposal is subject to special requirements in order to prevent infection	A
18 01 04	Wastes whose collection and disposal is not subject to special requirements in order to prevent infection (e.g. dressings, plaster casts, linen, disposable clothing, nappies)	
18 01 06*	Chemicals consisting of or containing dangerous substances	M
18 01 07	Chemicals other than those mentioned in 18 01 06	
18 01 08*	Cytotoxic and cytostatic medicines	A
18 01 09	Medicines other than those mentioned in 18 01 08	
18 01 10*	Amalgam waste from dental care	A

Any waste whose six-digit code is marked with an asterisk (*) is a hazardous waste. Classification may be absolute (A), defining waste as hazardous regardless of the concentration of any 'dangerous substance' within it, or a 'mirror entry' (M), covering wastes having the potential to be hazardous or non-hazardous depending on their composition and the concentration of 'dangerous substances' within them. The hazard potential is determined by reference to published threshold limits or, for infection hazards, on risk assessment.

2 Direct alcohol fuel cell

Fuel cells generate electric power from an electrochemical process using fuels such as hydrogen or methanol. Compared with batteries, fuel cells typically have a higher energy density and a lower weight. In addition, fuel cells are environment-friendly (especially if the fuel is taken from a renewable resource) and can be recharged instantly. Fuel cells are being used in prototype applications to power vehicles, cellular telephones, homes, commercial properties, laptops, household appliances and industrial machinery. A direct alcohol fuel cell (DAFC) uses liquid alcohols as a fuel and is very attractive as power source for mobile, stationary and portable applications. However, alcohols are very difficult to electro-oxidize completely and up to now methanol has been considered the most promising organic fuel. Carbon-supported PtRu nanoparticles (PtRu/C) are the best electrocatalysts for direct methanol fuel cell (DMFC), however, the synthesis of highly dispersed carbon supported PtRu nanoparticles with high loading remains a challenge. The conventional methods of preparation, like wet impregnation and reduction, do not provide satisfactory control of the particle size and distribution.

3 Waste segregation and recovery for DAFC supply

Further segregation is proposed as an effective method of reducing costs in waste disposal, and for achieving compliance with the new legislative controls that prohibit mixing of wastes [9]. Directed principally at the elimination of packaging wastes from the more costly sanitary waste stream, space constraints in clinical areas can present practical problems that limit the options for additional segregation. Though attractive on environmental, ecological and administrative grounds, such complex and comprehensive waste management procedures are likely to succeed only in newly-build hospitals where sufficient space is devoted to these core functions at the design and construction stages. Since space is at a premium in the majority of existing building stock, the elimination of packaging waste from clinical wastes might best be achieved by removal at source, in the supply department, though this requires care to protect the integrity of sterile supplies. In circumstances where segregation of wastes is undertaken solely at the point of arising, in the busy and often cramped hospital ward, the risks associated with incorrect segregation that might result in clinical wastes entering a more general domestic waste stream cannot be dismissed. Efficient supply chain management may thus offer an effective response to demands for the improved segregation of wastes, reducing costs without compromise of safety, and supporting effective risk reduction. However, based on the evidence reported here, it is apparent that there has been little if any improvement and that much still remains to be done to improve the standards of clinical waste management in hospitals.

Amid the aforementioned sanitary wastes, solvents containing alcohol are very interesting as fuel for DAFC. So, in hospital facilities, there is a need of creating specific areas for stocking wastes in a segregated way, especially for solvents containing alcohol (ethanol and ethylene glycol). This procedure would be permitted by some legislation, for instance, the European directive 2006/12/EC and the Italian one [10]. Ethanol and ethylene glycol are used in hospital facilities and they are good energetic vectors with respect to hydrogen and to hydrocarbons. They have a high solubility in water and energy density (*We*) like gasoline (10-11 kWh kg^{-1}) [11]. Moreover, since energy efficiency (ε_r) of DAFC is greater than that of hydrogen/oxygen-based cell fuel (table 2), that is, 83% at 25°C, this is the interesting reason for recovering ethanol and ethylene glycol from sanitary liquids for energetic objectives. Ethanol and ethylene glycol are good sources of energy, even if it is necessary for their electrochemical reaction to entail the increasing of anode overvoltage towards high values with respect to those reached by hydrogen (fig. 1). This inconvenience could be overcome by allowing the reaction to be performed on binary or ternary catalyzer based on platinum which allows the production of molecules with small dimensions that are easily oxidizable to CO_2 [12].

Table 2: Thermodynamic data of related to electrochemical oxydation of some alcohols.

Fuel	ΔG° /kJ mol^{-1}	E° /V vs SHE	E_{emf} /V	ΔG /kJ mol^{-1}	W_e /kWh kg^{-1}	ΔH /kJ mol^{-1}	ε_r
CH_3OH	−9.3	0.016	1.213	−702.7	6.09	−726.7	0.967
C_2H_5OH	−97.3	0.084	1.145	−1326.7	8.01	−1367.9	0.970
CH_2OH–CH_2OH	−4.8	0.005	1.224	−1181.5	5.29	−1190.2	0.992
C_3H_7OH	−168	0.097	1.132	−1965.3	9.10	−2023.2	0.971
C_4H_9OH	−243	0.105	1.124	−2602.1	9.77	−2676.8	0.972

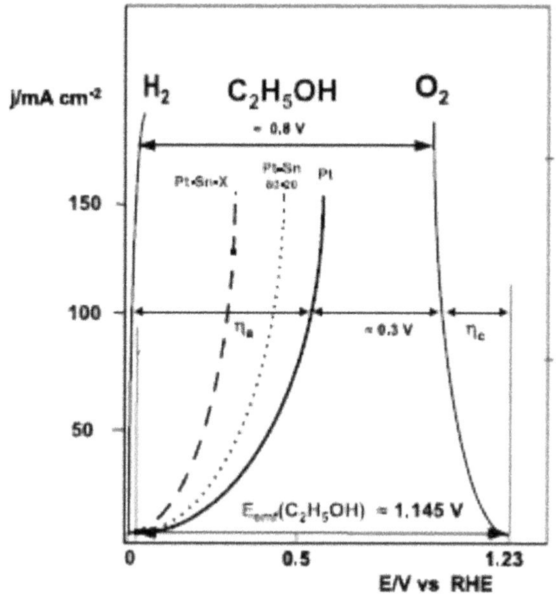

Figure 1: Comparison between voltage-current of a H_2/O_2-based cell fuel with platinum electrodes and DAFC ethanol with anodic ternary catalyzer (Pt-Sn-X).

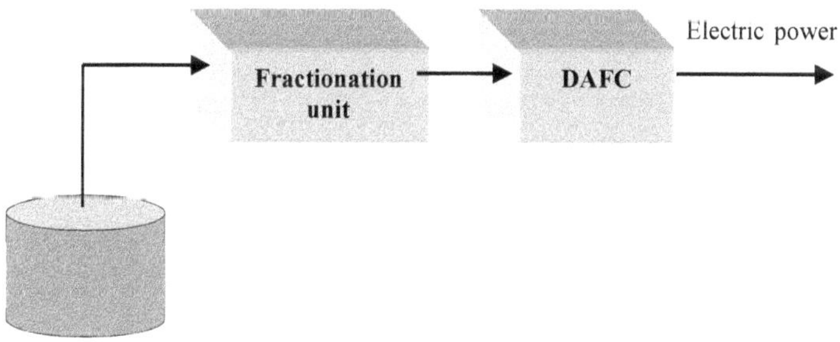

Figure 2: Block scheme of proposed architecture.

Fig. 2 illustrates the proposed architecture of processing segregated liquid wastes extracted from sanitary waste and from sanitary decayed radioactive wastes. Both types of waste, including other solvents that have not been segregated before, are stocked in a specific tank of fig. 2. Afterwards, the liquids are treated in a fractionation unit according to the plant of fig. 3.

Thanks to Antoine equation [13], using the appropriate constants, it is possible to calculate the vapor tension of both alcohols (ethanol and ethylene glicol), then, their volatility and the relative one.

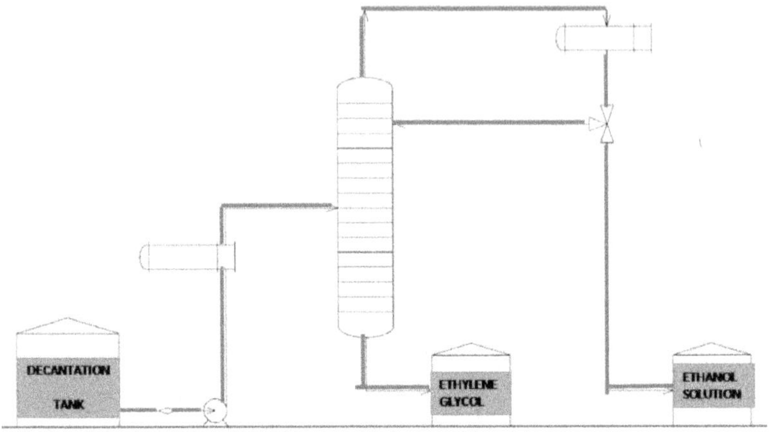

Figure 3: Fractionation plant/unit.

To calculate the number of steps necessary to obtain the destre separation, it is appropriate to use the FUG (Fenske-Underwood-Gilliland) method according to the following procedure:

$$N_{min} = \frac{\log\left(\frac{x_{D_i}}{x_{B_i}} \cdot \frac{x_{B_j}}{x_{D_j}}\right)}{\log \alpha^*} \tag{1}$$

$$1 - q = \Sigma \frac{\alpha_i x_{F_i}}{\alpha_i - \Theta} \tag{2}$$

$$R_{min} + 1 = \Sigma \frac{a_i x_{D_i}}{a_i - Q} \tag{3}$$

$$\frac{N - N_{min}}{N + 1} = 1 - \exp\left[\frac{(1 + 54.4X)(X - 1)}{(11 + 117.2X)\sqrt{X}}\right] \tag{4}$$

where X is given by

$$X = \frac{R - R_{min}}{R + 1} \tag{5}$$

and all symbols are explained under Nomenclature.

Finally, to establish in which stage it is better to feed, the Kirkbride equation has been used in an empirical way, that is,

$$\frac{N_R}{N_S} = \left[\left(\frac{B}{D}\right)\left(\frac{x_{F,HK}}{x_{F,LK}}\right)\left(\frac{x_{B,LK}}{x_{D,HK}}\right)\right]^{0.206} \quad (6)$$

The FUG method has allowed the determination of the following parameters for the design of the system:

$N_{min} = 2$, $R_{min} = 0.24$, $N = 7$, $R = 0.31$, number of feeding stages = 5.

4 Summary and conclusions

This research has illustrated the opportunity of recovering alcohols from sanitary wastes in order to use them for feeding a DAFC. In general, even if a segregation is carried out in clinical and hospital facilities, solvents containing alcohols are used in these facilities for different reasons. But an over-segregation that allows the recovery of alcohols from conventional sanitary wastes and decayed radioactive ones will produce great benefits for saving money and protecting the environment. The use of DAFC fuelled by recovered alcohols is essential for hospital pavilions like neonatology, resuscitation and surgery rooms, where, the electric energy must not be interruptible for health and safety reasons. The proposed research agrees with European and national legislations.

Nomenclature

B	Residual flow in kmol/h
D	Residual flow in kmol/h
F	Flow for tower in kmol/h
F_w	Flow for fractionation tower in kg/h
N	Effective number of stage for separation
N_{min}	Minimal number of necessary stages for separation
N_R	Number of stages over feeding
N_S	Number of stages under feeding
P	Operating pressure
$PM_{ethanol}$	Ethanol molecular weight
PM_{glycol}	Ethylene Glycol molecular weight
PM_{mix}	Mixed feeding molecular weight
$p°$	Vapor tension
R	Reflux
R_{min}	Minimal reflux
T_{eb}	Ebullition temperature
x_B	Fraction in moles of ethanol in residual
x_D	Fraction in moles of ethanol in the distilled
x_{Dw}	Fraction in weight of ethanol in the distilled

x_F	Fraction in moles of ethanol in the feeding
x_{Fw}	Fraction in weight of ethanol
α	Viscosity
α^*	Relative viscosity

References

[1] Mercier, C. & Ellam, T., Waste disposal. Waste not, want not. *Health Service Journal*, **106(5497)**, p. 27, 1996.

[2] Department of the Environment, Transport and the Regions (DETR), A better quality of life: A strategy for sustainable development for the United Kingdom, HMSO: London, 1999.

[3] Health Services Advisory Committee (HSAC), Safe Disposal of Clinical Waste, 2nd ed., HSE Books, The Stationery Office, 1999.

[4] Audit Commission, 1997b, Pembrokeshire NHS Trust, 1996.

[5] Act of 13 July 1992.

[6] Italian act DPR 254/2003.

[7] Organisation Mondiale de la Santé, la gestion des déchets des hopitaux et autres établissements de soins de santé, Rapport sur une reunion de l'OMS, Bergen 28 juillet – 1 juin 1983, Danemark, ISBN 9289022639.

[8] Muhlich, M., Scherrer, M. & Daschner, F.D., Comparison of infectious waste management in European hospitals. *Journal of Hospital Infection*, **55**, pp. 260-268, 2003.

[9] Townend, W.K. & Cheeseman, C.R., Guidelines for the evaluation and assessment of the sustainable use of resources and of wastes management at healthcare facilities. *Waste Manage Research*, **23**, pp. 398-408, 2005.

[10] Italian act Dlgs.152/06 Parte IV, art.179 s.m.i.

[11] Lamy, C., Lima, A., LeRhun, V., Delime, F., Coutanceau, C. & Léger, J.-M., Recent advances in the development of direct alcohol fuel cells (DAFC). *Journal of Power Sources*, **105(2)**, pp. 283-296, 2002.

[12] Laborde, H., Rezzouk, A., Léger, J.-M. & Lamy, C., Effect of the electrodestructure on the electrocatalytic oxidation of low-weight alcohols. Applications to the direct alcohol fuel cell. *Proc. of the Symposium on Electrode Material and Processes for Energy Conversion and Storage*, eds. S. Srinivasan, D.D. McDonald & A.C. Khandkar, The Electrochemical Society, pp. 275-294, 1994.

[13] Sinnott, R.K., *Coulson & Richardson's Chemical Engineering*, vol. 6: Chemical Engineering Design, 3rd ed., Butterworth-Heinemann: Oxford, 2003.

Gasified residual/waste biomass as solid oxide fuel cell feed for renewable electricity production

J. Jurewicz & N. Abatzoglou
Université de Sherbrooke, Canada

Abstract

Gasification of waste biomass, a renewable feedstock, is considered as a sustainable technology for producing environmentally friendly feed (biosyngas) for fuel cells. This work examines the contaminants contents (i.e. sulphur and halogens) of typical waste biomass sources and evaluates their nuisance to Solid Oxide Fuel Cells (SOFC) performance for the two most considered types of anodes: metal-ceramic composites (cermets) and composite oxides. The necessity and the extent of physico-chemical purification of the bio syngas and its cost depend upon the chemical stability of the anode material. These needs as well as their associated costs are evaluated and a first assessment of the sustainability of such solutions is undertaken and commented.
Keywords: waste, biomass, gasification, sulphur, halogens, SOFC, syngas, gas purification, anode, cermet, composite oxide ceramics.

1 Introduction

As world energy consumption increases (currently by ≈ 2% per annum [1]), the analysis [2] of energy scenarios concluded that it is possible to address simultaneously various sustainable development objectives, using available resources and technical options. The foreseen solutions are based on some combination of better exploitation of renewable resources, higher energy conversion efficiencies, and advanced energy conversion technologies. The latter includes the development of a variety of fuel cells, now widely identified [2] as a long term solution for higher than conventional energy conversion to electricity, co-generation (combined heat and power or CHP), motive power and transport. Some important characteristics of the studied fuel cell technologies are presented in table 1.

Table 1: Comparison of fuel cell technologies [3].

Fuel cell type	Operating temperature	System output	Electrical efficiency	CHP efficiency
	[°C]	[kW]	[%]	[%]
Polymer Electrolyte Membrane (PEM)	50–100	< 1–250	53–58 (transportation) 25–35 (stationary)	70–90 (low grade waste heat)
Alkaline (AFC)	90–100	10–100	60	< 80 (low grade waste heat)
Phosphoric Acid (PAFC)	150–200	50–1000	> 40	< 85
Molten Carbonate (MCFC)	600–700	< 1–1000	45–47	< 80
Solid Oxide (SOFC)	600–1000	< 1–3000	35–43	< 90

The high operating temperatures of molten carbonate (MCFC) and solid oxide fuel cells (SOFC) make them well suited for co-generation [4]. Moreover, SOFC technology offers considerably higher CHP efficiencies, potentially longer operating lifetimes [5], safer operation due to the absence of a molten solid phase and lower capital and operational costs than MCFC.

The typical "sandwich" type SOFC assemblies are composed of three, functionally distinct, layers (anode – electrolyte – cathode) where the anode is designed for electrochemical oxidation of the fuel associated with the charge transfer to a conducting contact. The most critical characteristics of materials considered as potential candidates for SOFC anodes are: electronic conductivity; oxygen diffusivity (ionic conductivity); oxygen surface exchange (reactivity); chemical stability and compatibility; thermal expansion; mechanical strength and dimensional stability under redox cycling.

The particular advantage of the SOFC concept over all other types of fuel cells, is its ability to operate with a variety of fuels, both gaseous – such as methane, biogas, biosyngas, and liquids – such as gasoline, jet fuel, diesel fuels and oxygenates, e.g. methanol, ethanol (and bio-ethanol) and "green diesel". However such a variety of fuels with their inherent contaminants makes more stringent the requirements of anode's chemical stability and compatibility.

This capacity of solid oxide fuel cells to work with fuels originating from renewable resources like biomass is of particular interest.

The aim of this work is to detail the constraints imposed in order to render compatible the use of syngas originating from residual and waste biomass gasification with the SOFC anodes.

2 Residual biomass availability

While renewable, so called "Green Power", resources (hydro, wind, solar thermal and photovoltaic, geothermal, marine and biomass) are geographically more evenly distributed than fossil and nuclear resources, and energy flows from renewable resources are several thousand times higher than current total global energy use, the statistics [2] indicate that only 13.7% of worldwide primary energy use comes from renewable resources. The observed [6] most recent increase of global trade of woody biomass, primarily for energy, from 5.6 million tons in 2003 to over 11 million tons in 2007, witnesses the interest in such source of green energy. The economic potential of renewable sources is affected by many constrains including competing land use, the amount and timing of solar irradiation and wind patterns, as well as some other environmental factors. The land use is to be given priority for both human and animal feeding and, in most cases where the soil allows it, even the residuals from crop and forest harvesting are preferably to be left to replenish land's nutrients [7]. However it is proposed [6] to periodically harvest for long-lived wood products which results in substantially greater carbon storage than when forests are left in an unmanaged state. The US Energy Information Administration, which forecasts US energy production, considers four components to the biomass supply schedule: agricultural residues, energy crops, forestry residues and urban wood waste/mill residues. The analysis [8] of their price and availability reveals that the less costly (1.42 $/GJ) is almost exclusively post-consumer urban wood waste/mill residue. Next (at 1.90 $/GJ and higher) comes the agricultural residues followed by energy crops and forestry residues (at 2.18 $/GJ or higher). Such results confirmed similar projections obtained earlier [9].

The post-consumer urban wood waste/mill residues include:

- wood materials (like slabs, edgings, trimmings, sawdust, veneer clippings and cores, and pulp screening) and bark generated at primary (lumber) manufacturing plant;
- similar wood scraps and sawdust from woodworking shops, furniture factories, wood container and pallet mills;
- wood residues from municipal solid waste (wood chips and pallets), utility tree trimmings and construction and demolition sites.

In the European market [10], the post-consumer wood is classified as:

- A: quality wood: clean wood;
- B: quality wood: slightly contaminated, e.g. with paints, glues and coatings;
- C: quality wood: hazardous wood waste contaminated with heavy metals, fire retardants and wood preservatives.

The analysis [11] of markets for recovered wood in Europe indicates, that in Germany only in year 2003 up to 6.531 million tonnes of post-consumer wood were available for the price of 20–30 euro/ton air dry of A – quality wood and 7–17 euro/ton air dry for B – quality wood.

Such variety of wood residue feedstock is usually mechanically homogenized and converted into commercially available product – wood pellets.

3 Wood pellets as a fuel

Wood pellets as heating fuel originated in the US during the 1970s in response to high energy prices [12] but became recently a significant form of biomass consumption for energetic purposes with world market trades exceeding 3,000,000 tonnes per year [13]. Quite thorough analyses of global wood pellets markets and industry including policy drivers, market status and raw materials has been published recently [14]. Sweden, Canada and USA are world leading pellets producers with an annual production capacity exceeding 3,500,000 tonnes [14].

The production of wood pellets is based on several mechanical operations of milling interlaced with drying. This sequence is ending up with pressing through a mould with simultaneous plasticizing (due to lignin and resin content) of the wood. The latter operation may require some additives (like starch) for sawdust originating from hardwood (less rich in lignin) feedstock for preservation of the final shape of the pellet.

The chemical elemental composition of such biomass-derived fuels depends upon the biomass type [15], the origin of the residue and its pre-processing prior to its final destination (use). The fuel components, once consumed, will all end up transformed preferably to element's oxides either in the gaseous (i.e. CO_2) or in solid form (ashes) and will be released to environment. While carbon dioxide and water are neutral, the release of all other compounds to the environment is regulated by norms; namely by standardization of the biomass fuels initial composition. Table 2 presents the acceptable limits of sulphur and chlorine, both elements being considered as hindering the long term operation of SOFC devices.

Table 2: Tolerated limits of sulphur and chlorine according to prevailing worldwide legislation.

	Element:	Sulphur	Chlorine
Austrian ÖNORME M7135	Wood pellets:	≤ 0.04 %*	≤ 0.02 %
	Bark pellets:	≤ 0.08 %*	≤ 0.04 %
German DIN51731/DIN plus		≤ 0.08 %	≤ 0.03 %
Swedish SS 187120	Group 1:	≤ 0.08 %	≤ 0.03 %
	Group 2:	≤ 0.08 %	≤ 0.03 %
	Group 3:		
CEN (draft) CEN/TS 14961: 2005 Annex A		≤ 0.05 %	**

* Dry basis;
** Recommended to be stated in category: Cl 0.03; Cl 0.07; Cl 0.10; Cl 0.10 + (if Cl > 0.10% the actual value to be stated).

The comparison of the sulphur limits, as imposed by different legislations, with residual sulphur contents in a variety of clean (pristine – without any man-made additives) biomass [16] indicates, that pure bark pellets, as well as many herbaceous fuels, could not be accepted – their residual sulphur content being above the imposed limits.

The combustion of contaminated (B – quality) wood pellets for heating purposes bears some environmental consequences like increase of particulate emissions [17] beyond the current EU emission limit for dust as well as formation of dioxins [18].

4 Wood pellets from Canadian sources and their life cycle analysis

Canada has substantial Green Power resources and the natural potential to generate about half of its current electricity needs (~590 TWh) using Green Power [19]. While current installed, biomass based technologies, count for 1,935 MW (just above 1% of its total generation capacity) percentage substantial rise might be expected when co-generation technologies will be used [19].

The Canadian wood pellet industry is predominantly located in 3 provinces: British Columbia (33% of production), Quebec (28% of production) and Ontario (20% of production) [20]. Its production capacity as well as different market sales are presented in fig. 1.

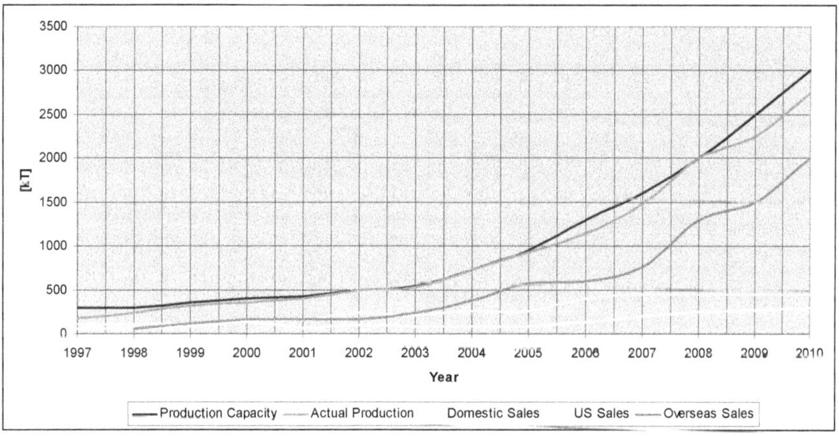

Figure 1: Evolution of Canadian wood pellet industry and different markets [21].

The origins of the raw materials for wood pellets production are the following:

- Saw mills; where harvested wood (pine, Douglas fir, spruce, birch and Western Red Cedar) is converted into lumber (construction wood) and remaining saw dust, bark and other residues like shavings, standing for about 15% [2] of the total volume of the harvested wood, are transformed into pellets.

- Standing dead wood either after a forest fire or as a result of infestation (e.g. infestation of pine forest by Mountain Pine Beetle). Both reasons are making remaining wood unsuitable for usual application in timber or paper industries. Such raw material (whole tree trunk) requires more energy to convert it into the shavings size needed by pellet industry.

Such sources might be extended significantly by harvesting forest thinning, which are considered not only unnecessary to replenish forest nutrients but also the main reason for rapid spreading of forest fires. According to Canadian Forest Service [22] between 2.5 to 3 million hectares of forests are lost to about 10,000 forest fires every year.

Canada is the world's most prominent exporter of wood pellets and its main partner is Europe. Such long haul transport, means an energy consumption (mainly fossil) estimated [23] at 2.40 $/GJ for a 35,000 ton vessel travelling a distance of 18,000 km [24]. Moreover, it represents a high environmental nuisance because of the Green-House-effect Gases (GHG) generated during the overseas transport [25]. The life-cycle assessments of Canadian wood pellets exported from the province of British Columbia (Prince George) to Sweden (port of Stockholm) are presented in table 3.

Table 3: Energy consumed for each stage of the wood pellet production and transport (adapted from [26]).

Parameter	Process stage					
	Harvesting of lumber	Transport*	Production: Sawdust as fuel	Production: Natural Gas as fuel	Transport by train (750 km)	Transport by ocean vessel (15500 km)
Energy consumed [GJ/ton]	0.52	0.07	3.78	2.97	0.26	2.60
Energy consumed if sawdust as fuel [%]	7	1	52	-	4	36
Energy consumed if Natural Gas as fuel [%]	8	1	-	46	4	40
*Trucking of lumber to mill (110 km) and residues to pellet plant (27 km)						

The analysis of above data suggests, that while half of the energy is consumed for actual pellets production, their post-production transport requires between 40 and 44% of total energy.

5 Gasified biomass as a feed for SOFC devices

Gasification appears to be the dominant biomass conversion technology. The gaseous products (syngas or biosyngas) from gasification processes are considered as vectors of high-efficiency power production or of the synthesis of chemicals and fuels. Each application requires specific syngas composition and the existing gasification technologies offer this possibility because this composition depends upon the gasification process, the gasification agent (air, oxygen, oxygen-enriched air and their mixtures with steam) and the processing temperature. Generally [27], for gasification processes carried-out at high temperatures (above 1200°C) the biomass is completely converted into CO and H_2 (besides H_2O and CO_2) with an absence of tars which makes this technology interesting for a variety of chemical synthesis routes leading to high-end products. For processing temperatures between 800–1000°C, the as-produced gas composition is characterized by a larger variety of chemical species including (besides CO_2 and H_2O) CO, H_2, CH_4, aliphatic hydrocarbons, benzene, toluene and heavier compounds under the generic name "tars". A thorough review on tar is available in the literature [28]. The energy contained in typical syngas components (CO, H_2) stands for about 50% of the total, the remaining 50% being carried by CH_4 and higher (aliphatic and aromatic) hydrocarbons making such gas useful for power generation purposes or for production of synthetic natural gas (SNG) [27]. Such as-obtained gas requires important post-gasification treatment like thermal cracking or reforming to match the quality of the gas produced during gasification at higher temperatures.

Among the variety of gasification technologies classified as direct (auto-thermal) and indirect (or allo-thermal) processes the latter group is of particular interest as it offers nitrogen-free product gas as no oxygen is required for the gasification as well as carbon-free ashes. Another advantage of this technology is the possibility of operating it at smaller scale (less than 5 MW_{th}) making it interesting candidate for distributed generation devices where the produced gas is fed into solid oxide fuel cells unit. During the analysis of such an option it is important, among others, to compare the produced gas composition with that required by SOFC unit. The composition of the gas produced from wood by three different indirect gasification technologies is presented in table 4.

An analysis of the gas composition reveals that all combustibles contained in the produced gas are "consumable" at typical, Ni-based cermet type SOFC's anode:

- Both hydrogen and carbon monoxide are desired SOFC fuels;
- Methane, C_2+ hydrocarbons and benzene are being reformed to CO and H_2 at high temperature over such anodes with catalytic reforming activity in the presence of steam originating from hydrogen consumption over anode's surface;
- Ammonia is considered [29–31] as very interesting SOFC fuel;
- Carbon deposition over anode's surface due to tar presence may be avoided under specific current density [32];
- Solid particles are filtered [33] from the gaseous fuel prior to SOFC unit.

Table 4: Typical gas composition of three indirect gasification processes [27].

Gas component, dry basis	Process		
	1	2	3
Hydrogen H_2 [vol %]	30–45	20–22	15–20
Carbon monoxide CO [vol %]	20–30	41–44	40–43
Carbon dioxide CO_2 [vol %]	15–25	11–14	10–12
Methane CH_4 [vol %]	8–12	12–16	15–17
C_2+ hydrocarbons [vol %]	1–3	4–6	5–6
Benzene C_6H_6 [vol %]		1	1
Nitrogen N_2 [vol %]	1–3	2–10	1–4
Ammonia NH_3 [ppmV]	500–1000		500–1000
Hydrogen sulphide H_2S [ppmV]	50–120		40–100
Tar [g/m_n^3]	0.5–1.5	40	40
Particles [g/m_n^3]	10–20		-

1 – Fast Internal Circulation Fluidised Bed (FCIB) process;
2 – SilvaGas process;
3 – MILENA process.

The only component of biomass gasification product, which is undesired, is the hydrogen sulphide. At very low concentrations of 0.05; 0.5 and 2 ppm, tested at 1023, 1173 and 1273 K, respectively, the anode activity losses are reported [34] to be reversible once the fuel no longer contains H_2S. At temperatures below 1073 K the sulphur poisoning was found [35] to be irreversible. The cell voltage drop increases significantly with H_2S concentration in the fuel from 1 to 3 ppm and saturates at 5 ppm [36]. The reforming of methane on catalytically active sites is affected [37] as well.

To prevent such decrease in Nickel cermet based SOFC performances the removal of sulphur bearing compounds from the feed gas to acceptable level is required. A recent thorough review on H_2S removal technologies is available in the literature [38]. Although this review covers the biogas purification most of the reviewed technologies can be used as well in the case of the syngases. The extrapolation of the permissible sulphur concentration in the natural gas feed, considering a deactivation rate of Ni-based planar SOFC of 0.75% per 1000 h (6.4% per year) has indicated the value of 18 ppb [35]. Such "deep" cleaning is very difficult to reach; only zinc-based sorbents operating at 673–823 K are reported [39] to be effective in removing of H_2S to 1 ppm levels at elevated temperatures but not high enough to match typical operating temperatures of gasifier/SOFC tandem. The known deep gas cleaning technologies offer the possibility to lower the residual sulphur content but with excessive cost making such operation much less profitable from the economic stand point. The reported [40] cost of fuel processing module (sulphur removal unit, catalyst, fuel preheater and ejector) accounts for 33% of the total (without stack) balance of power (BOP) investment for 5 kW unit for building applications. The model for 250 kW planar SOFC unit for CHP application specifies [41] the cost of fuel system as 8% where the stack counts for 31% of total cost.

It is clear that high sensitivity of Ni-cermet based anodes to sulphur bearing compounds inherently present even in pristine biomass makes the gas conditioning circuit more complex and increase overall investment and operational costs of BOP. The possible Nickel-free, sulphur-tolerant replacement for SOFC anode materials like fluorite, rutile, tungsten bronze, pyrochlore, perovskite and spinel structures were indentified [42] and reviewed [43–45]. The direct oxidation of methane [46, 47] or diesel fuel [48] over substituted perovskites anodes are reported as well. The all-perovskite (anode, electrolyte and cathode) SOFC may be of particular interest due to structural similarity of layers. Such fuel cell [49], based on the "sandwich" of the $(La_{0.75}Sr_{0.25})_{0.95}Cr_{0.5}Mn_{0.5}O_{3-\delta}$ (LSCM) as the anode, $La_{0.8}Sr_{0.2}Ga_{0.8}Mg_{0.15}Co_{0.05}O_{3-\delta}$ (LSGMCo) as the electrolyte and $Gd_{0.4}Sr_{0,6}CoO_{3-\delta}$ (GSC) as the cathode, can minimize the polarization losses between electrolyte and electrodes, where the slight inter-diffusion between the perovskite components observed results in better contact at the interface, allowing for smoother transfer of oxygen ions between the electrodes and electrolyte.

However the actual performances of composite oxide anodes do not match these obtained for Ni-based cermet anodes. More research effort has to be made for optimization of both their composition and performances. For such complex structures the application of the combinatorial chemistry approach in their synthesis is of particular interest.

Another concern for more widespread SOFC usage is the cost of their fabrication. The thermal plasma spraying, which is industrially proven and widely accepted technology, offers both rapid prototyping through combinatorial chemistry method as well as less-costly industrial mass-production allowing integrated deposition of all subsequent layers (anode, electrolyte and cathode) in one operation.

6 Conclusion

The possibility of significant increase in biomass energy conversion efficiency for SOFC based Combined Heat and Power generation cycle is of special interest especially for colder climate regions like northern Canada.

The export of Canadian biomass as far as Europe makes overall energy balance (energy contained in biomass vs energy consumed for its transport) less favourable – up to 40 % of energy is lost to transport.

The gasification process, compared to simple combustion or incineration, offers a better control of the release to environment of undesired conversion products, like manmade impurities contained in lesser quality wood's residues, while still allowing for the full usage of their energy content.

The direct gasification of as-obtained biomass (without drying, size reduction and pelletizing stages) may preserve both energy spent for pellets formation and that contained in the volatiles (mainly terpenes) usually lost [50] during drying process.

The sulphur content represents a technological challenge for the gas conditioning technologies because Nickel-based SOFCs are ultra-sensitive to sulphur bearing compounds. Consequently, the cleaning is usually complex and costly.

The SOFC based on composite oxide anodes are sulphur tolerant and are considered as a replacement to Ni cermet anodes; however their actual electrochemical performances are still inferior to the Nickel based ones.

The variety of the possible structures of composite oxides (perovskite-related like titanates, chromites, vanadates, gallates, niobates and ferrites; cubic fluorite related like zirconia-based and ceria-based as well as pyrochlores and spinels) makes combinatorial chemistry useful promising tool towards optimizing SOFC anode's elemental composition and structure.

The thermal plasma deposition technology allows for rapid SOFC prototyping using combinatorial chemistry approach as well as further industrial mass production of entire multi-layered fuel cells.

References

[1] Mason, J.E., World energy analysis: H_2 now or later. *Energy Policy*, **35**, pp. 1315-1329, 2007.

[2] Goldemberg, J., Johansson, T.B. & Anderson, D., *World Energy Assessment, Overview 2004 Update*, United Nations Report, 2004.

[3] US Department of Energy.

[4] Williams, R.H., Advanced energy supply technologies. *World Energy Assessment*, United Nations Report, 1998.

[5] Bakker, W., Cohn, A. & Gloldstein, R., Advances in solid oxide fuel cells. *EPRI Journal*, **21(5)**, pp. 42-45, 1996.

[6] UNECE/FAO, Forest products annual market review, 2008.

[7] Pimentel, D., Moran, M.A., Fast, S., Weber, G., Bukantis, R., Balliett, L., Boveng, P., Cleveland, C., Hindman, S. & Young, M., Biomass energy from crop and forest residues. *Science*, **212**, pp. 1110-1115, 1981.

[8] Haq, Z. & Easterly, J.L., Agricultural residue availability in the United States. *Applied Biochemistry and Biotechnology*, **129-132**, pp. 3-21, 2006.

[9] Easterly, J.L. & Burnham, M., Overview of biomass and waste fuel resources for power production. *Biomass and Bioenergy*, **10(2-3)**, pp. 79-92, 1996.

[10] Leek, N.A., Data collection for post-consumer wood in the Netherlands, http://www.unece.org/fileadmin/DAM/timber/workshops/2008/wood-balance/presentations/08_Leek.pdf, Probos.

[11] Van Benthem, M., Leek, N., Mantau, U. & Weimar, H., Markets for recovered wood in Europe; case studies for the Netherlands and Germany based on the BioXchange project. *Proc. of the 3rd European COST E31 Conf.*, Klagenfurt, Austria, 2007.

[12] Kotrba, R., Closing the wood pellet gap. *Biomass Magazine*, February, 2009.

[13] Forest News WIRE, PR press release at: http://www.pr.com/press-release/89107, June, 2008.

[14] Peksa-Blanchard, M., Dolzan, P., Grassi, A., Heinimö, J., Junginger, H.M., Ranta, T.-M. & Walter, A., Global wood pellets markets and industry:

policy drivers, market status and raw material potential, IEA Bioenergy Task 40 report, November, 2007.
[15] Nordin, A., Chemical elemental characteristics of biomass fuels. *Biomass and Bioenergy*, **6(5)**, pp. 339-347, 1994.
[16] Obernberger, I., Brunner, T. & Bärnthaler, G., Chemical properties of solid biofuels – significance and impact. *Biomass and Bioenergy*, **30**, pp. 973-982, 2006.
[17] Khan, A., Aho, M., de Jong, W., Vainikka, P., Jansens, P.J. & Spliethoff, H., Scale-up study on combustibility and emission formation with two biomass fuels (B quality wood and pepper plant resudue) under BFB conditions. *Biomass and Bioenergy*, **32(12)**, pp. 1311-1321, 2008.
[18] Lavric, A.D., Konnov, A.A. & Ruyck, J.D., Dioxin levels in wood combustion – a review. *Biomass and Bioenergy*, **26**, pp. 115-145, 2004.
[19] Canadian Renewable Energy Alliance, Green Power for Electricity Generation – Creating an Industry in Canada, August, 2006.
[20] EUBioNet2, No 10-2.
[21] Adapted from Canadian Wood Pellet Industry web page.
[22] Natural Resources Canada, Forests, Fire, http://www.nrcan.gc.ca/forests/fire/13143, 2014.
[23] Aruna, P., Laarman, J.G., Araman, P. & Cubbage, F.W., An analysis of wood pellets for export: a case study of Sweden as an importer. *Forest Products Journal*, **47(6)**, pp. 49-52, 1997.
[24] Duffy, A. & Conroy, M., Embodied transport energy analysis of imported wood pellets. *Transactions on Ecology and the Environment, vol. 105, Energy and Sustainability*, eds. C.A. Brebbia & V. Popov, WIT Press: Southampton, pp. 299-305, 2007.
[25] Svedberg, U., Samuelsson, J. & Melin, S., Hazardous off-gassing of carbon monoxide and oxygen depletion during ocean transportation of wood pellets. *Annals of Occupational Hygiene*, **52(4)**, pp. 259-266, 2008.
[26] Magelli, F., Boucher, K., Bi, H.T., Melin, S. & Bonoli, A., An environmental impact assessment of exported wood pellets from Canada to Europe. *Biomass & Bioenergy*, **33**, pp. 434-441, 2009.
[27] Boerrigter, H. & Rauch, R., Review of applications of gases from biomass gasification, Rapport ECN.RX-06-066, 2006.
[28] Milne, T., Abatzogloy, N. & Evans, R.J., Biomass gasifier "tars": their nature, formation, and conversion. *The Biomass Energy Foundation Press*, ISBN 1-890607-14-2, 1998.
[29] Fournier, G.G.M., Cumming, I.W. & Hellgardt, K., High performance direct ammonia solid oxide fuel cell. *Journal of Power Sources*, **162**, pp. 198-206, 2006.
[30] Meng, G., Jiang, C., Ma, J., Ma, Q. & Liu, X., Comparative study on the performance of a SDC-based SOFC fuelled by ammonia and hydrogen. *Journal of Power Sources*, **173**, pp. 189-193, 2007.
[31] Fuerte, A., Valenzuela, R.X., Escudero, M.J. & Daza, L., Ammonia as efficient fuel for SOFC, *Journal of Power Sources*, **192(1)**, pp. 170-174, 2009.

[32] Singh, D., Hernández-Pacheco, E., Hutton, P.N., Patel, N. & Mann, M.D., Carbon deposition in an SOFC fuelled by tar-laden biomass gas: a thermodynamic analysis. *Journal of Power Sources*, **142**, pp. 194-199, 2005.

[33] Sharma, S.D., Dolan, M., Park, D., Morpeth, L., Ilyushechkin, A., McLennan, K., Harris, D.J. & Thambimuthu, K.V., A critical review of syngas cleaning technologies – fundamental limitations and practical problems. *Powder Technology*, **180**, pp. 115-121, 2008.

[34] Matsuzaki, Y. & Yasuda, I., The poisoning effect of sulphur-containing impurity gas on a SOFC anode: Part I. Dependence on temperature, time, and impurity concentration. *Solid State Ionics*, **132**, pp. 261-269, 2000.

[35] Cunningham, R.H., Fowles, M., Ormerod, R.M. & Staniforth, J., Sulphur poisoning of the active materials used in SOFCs, *Report to Rolls-Royce*, F/01/00222/REP, URN 04/559, 2004.

[36] Sasaki, K., Susuki, K., Iyoshi, A., Uchimura, M., Imamura, N., Kusaba, H., Teraoka, Y., Fuchino, H., Tsujimoto, K., Uchida, Y. & Jingo, N., H_2S poisoning of solid oxide fuel cells. *Journal of the Electrochemical Society*, **153(11)**, pp. A2023-A2029, 2006.

[37] Ouweltjes, J.P., Aravind, P.V., Woudstra, N. & Rietveld, G., Biosyngas utilization in solid oxide fuel cells with Ni/GDC anodes. *Journal of Fuel Cell Science & Technology*, **3(4)**, pp. 495-498, 2006.

[38] Abatzoglou, N. & Boivin, S., A review of biogas purification processes. *Biofuels, Bioproducts and Biorefining*, **3**, pp. 42-71, 2009.

[39] Aravind, P.V., Outweltjes, J.P., de Heer, E., Woudstra, N. & Rietveld, G., Impact of biosyngas and its components on SOFC anodes. *Electrochemical Society Proc*eedings, **2005-07**, pp. 1459-1467, 2005.

[40] Braun, R.J., Klein, S.A. & Reindl, D.T., Assessment of solid oxide fuel cells in building applications, *Solar Energy Laboratory*, University of Wisconsin Madison, prepared for: Energy Center of Wisconsin, 2001.

[41] Fontell, E., Kivisaari, T., Christiansen, N. & Hansen, J.B., Conceptual study of a 250 kW planar SOFC system for CHP application. *Journal of Power Sources*, **131**, pp. 49-56, 2004.

[42] Tao, S. & Irvine, J.T.S., Discovery and characterization of novel oxide anodes for solid oxide fuel cells. *The Chemical Record*, **4**, pp. 83-95, 2004.

[43] Gong, M., Liu, X., Trembly, J. & Johnson, C., Sulfur-tolerant anode materials for solid oxide fuel cell application. *Journal of Power Sources*, **168**, pp. 289-298, 2007.

[44] Sun, C. & Stimming, U., Recent anode advances in solid oxide fuel cells. *Journal of Power Sources*, **171**, pp. 247-260, 2007.

[45] Goodenough, J.B. & Huang, Y.H., Alternative anode materials for solid oxide fuel cells. *Journal of Power Sources*, **173**, pp. 1-10, 2007.

[46] Tu, H., Apfel, H. & Stimming, U., Performance of alternative oxide anodes for the electrochemical oxidation of hydrogen and methane in solid oxide fuel cells. *Fuel Cells*, **6(3-4)**, pp. 303-306, 2006.

[47] Tao, S.W., Irvine, J.T.S. & Plint, S.M., Methane oxidation at redox stable fuel cell electrode $La_{0,75}Sr_{0,25}Cr_{0,5}Mn_{0,5}O_{3-\delta}$. *The Journal of Physical Chemistry B*, **110**, pp. 21771-21776, 2006.

[48] Liu, D.J. & Krumpelt, M., Activity and structure of perovskites as diesel-reforming catalysts for solid oxide fuel cell. *International Journal of Applied Ceramic Technology*, **2(4)**, pp. 301-317, 2005.

[49] Tao, S.W., Irvine, J.T.S. & Kilner, J.A., An efficient solid oxide fuel cell based upon single-phase perovskites. *Advanced Materials*, **17**, pp. 1734-1737, 2005.

[50] Ståhl, M., Granström, K., Berghel, J. & Renström, R., Industrial processes for biomass drying and their effect on the quality properties of wood pellets. *Biomass and Bioenergy*, **27**, pp. 621-628, 2004.

Poultry litter valorization to energy

P. De Filippis, M. Scarsella, N. Verdone,
M. Zeppieri & B. de Caprariis
"La Sapienza" University of Rome, Italy

Abstract

Historically manure has found utilization as fertilizer in agriculture because it contributes to the fertility of the soil by adding organic matter and nutrients, such as nitrogen, phosphorus and potassium. However, the current European Directive 91/676 reduces drastically the application of this material as fertilizer due to its high nitrate content. Therefore, identification of alternative eco-friendly disposal routes with potential financial benefits has become the need, and particularly promising is the energetic valorization of these biomasses. Poultry litter represents one of the more challenging bio-fuel feedstock for energy generation, being easy to handle and showing a composition that potentially assures a high energy content and the production of a byproduct (ash) with good fertilizing properties. In this paper an evaluation of the technical and economic feasibility of the energy conversion technologies usable to recover both the potential energy and fertilizer properties of poultry litter was provided. The focus was on the poultry farms of the North-East Italy, where more than half of the national poultry production is concentrated. The preliminary cost analysis suggests that actually energy production from poultry litter is economically viable in the case of large off-site plants only by means of anaerobic digestion, while in the case of small plants operating in situ for a mean holding capacity farm, utilization of gasification appears to be an option.
Keywords: manure, poultry litter, energy recovery.

1 Introduction

Owing to the population's growth and incomes, along with changing food preferences, the demand for livestock products is continuously increasing, while globalization is boosting trade in livestock inputs and products. Global production of meat and milk is forecasted to double in 50 years from the beginning of this century, reaching 465 million tonnes for the meat and overcoming one thousand million tonnes for the milk [1].

The present and future impact of the livestock sector on environmental problems is massive and in this context the management of manure has become one of the most important issues. Historically manure has found utilization as fertilizer in agriculture because it contributes to the fertility of the soil by adding organic matter and nutrients, such as nitrogen, phosphorus and potassium. These disposal methods, however, require proper land application and agronomic rates and soil management to ensure beneficial use and avoid leaching of nitrate and possible transfer of pathogens to the groundwater from fields where high doses of manure have been applied and threatening of soil fertility owing to unbalanced or even noxious nutrients concentrations [2]. As a consequence, land spreading of manure is strictly regulated and in EU the European Directive 91/676 reduces drastically the application of this material as fertilizer.

Therefore, identification of alternative eco-friendly disposal routes with potential financial benefits has become the need. Schemes providing energy and easy to handle fertilizer as a by-product would be the best alternative route.

To this aim poultry litter represents one of the more challenging bio-fuel feedstock for energy generation, being easy to handle and showing a composition that potentially assures a high energy content and the production of a byproduct (ash) with good fertilizing properties.

Commercial poultry production is rapidly expanding growing at 4–5% per year [3, 4] and is expected to continue expansion in the foreseeable future. Nowadays the world magnitude of the poultry industry has reached a total of all types of poultry meat productions of about 86 million ton per year [5]. Increases in the demand of poultry products have led to a growth in the dimensions of poultry farms and to a concentration of production, which has caused excessive manure supplies in certain areas.

Poultry industry is very developed in Italy with a poultry meat production over one million tons per year and an egg production of about 13 billion per year [6]. Moreover, most of the Italian poultry farms are concentrated in a relatively limited part of the national territory. In these areas, a mean holding capacity of about 40000 birds is calculated for the poultry companies, thus giving rise to an urgent need of robust and economical high-volume alternative uses for poultry litter.

2 Poultry litter material

Poultry facilities produce litter which is composed primarily of manure, bedding material (wood shavings, saw dust, paper etc.) and feed. Numerous factors determine the quality and composition of the litter, including the used bedding material, the number and type of poultry, the number of flocks per year, the sources of moisture, other than the overall litter management approach. However, whatever the litter composition is, except ash and water the other constituents are all potential energetic materials and poultry litter can be used for power generation directly. Moreover, no fuel preparation is required for poultry litter, except possibly some breaking up of clumps.

In the present work, two different kind of poultry litter coming from the poultry farms of the North-East Italy, where more than half of the national poultry production is concentrated, were considered. As can be noticed from the values reported in the Table 1, this feedstock presents a high variability of composition that mainly depends on the different materials used for the bedding.

After collection of representative samples, the poultry litter samples were dried in the shadow for one week before characterization. The results of the characterization analysis are reported in table 1.

Table 1: Main characteristics of the considered poultry litter.

	Poultry litter 1	Poultry litter 2
Moisture (mass %)	20.2	12.5
Volatile substance (mass % dry basis)	59.5	61.4
Fixed carbon (mass % dry basis)	24.9	11.2
Ash (mass % dry basis)	15.6	27.4
LHV (MJ kg^{-1})	12.72	12.68
Element (mass % dry basis)		
Carbon	37.2	35.9
Hydrogen	5.52	4.98
Oxygen	23.0	27.38
Nitrogen	4.4	3.56
Sulfur	0.3	0.77
Chlorine	0.8	0

The ash is characterized by an amount of phosphorus as P_2O_5 and potassium as K_2O of 20.2% and 13.8%, respectively.

3 Energy conversion technologies

Three technologies are currently tested as alternatives to land-disposal methods: anaerobic digestion, direct combustion and gasification.

Anaerobic Digestion: it is a biological process where microorganisms convert organic materials to methane, carbon dioxide and other organic compounds in the absence of oxygen [7]. Anaerobic digesters typically consist of large fermentation tanks with mechanical mixing, heating and gas collection. The process can be applied at any site that produces organic materials. The efficiency of conversion of manure to methane gas depends on many factors such as quality of manure, retention time in digester and temperature of the digester [8]. The retention time is strictly dependent on the operating temperature. When the temperature is in the mesophilic range (20–45°C), the typical retention time is 15 to 30 days, however, operating the process with temperature in the thermophilic range (temperature > 45°C), the retention time can be shortened to 12 to 14 days.

Thermophilic digestion systems offer higher methane production and better virus and pathogen reduction, but require greater energy input and higher cost of operation compared to mesophilic systems [9].

In evaluating the anaerobic digestion, the amount of volatile matter is of great importance, because typically only the 60% of volatile matter is converted to biogas [9] producing about 0.4 $m^3_{gas}/kg_{volatile\ matter}$. Depending on the waste feedstock and the system design, biogas is typically 55–75% methane. The produced biogas can be used to fuel internal combustion engines that run generators and produce electricity.

When poultry litter is considered, on the basis of its composition an estimated production of 0.22 m^3/kg_{litter} of biogas is obtained with a typical heating value of 22.4 MJ/m^3. The net electric power generation per MJ of thermal power depends on the method by which the electricity is generated. When a reciprocal engine was considered, a 30% efficiency conversion was assumed. With these assumptions the net electric power generation is 0.41 kWh/kg_{litter}.

The cost of an anaerobic-digestion system can vary dramatically depending on its size, intended purposes and sophistication. From the quotation of a European supplier for different anaerobic digester sizes (10, 100 and 1000 kWe) a capital cost equal to $11780 \times (kWe_{installed})^{0.86}$ is obtained, while the running costs are estimated from the same supplier to be $2600 \times (kWe_{installed})^{0.53}$ per year.

Direct Combustion: direct combustion is the simplest and most versatile energy conversion technology. Various fuels could be used including different varieties of biomass (e.g., wood, animal manure, agricultural residue and municipal solid waste). It involves the burning of fuel with excess air, producing hot flue gases that are used to produce steam in the heat exchange sections of boilers. Steam is then used to produce electricity in steam turbine generators.

On the basis of biomass combustion plants running in Italy, the conversion efficiency from thermal to electric power is assumed to be 21% for small plant and 32% for a plant of 10 MW or higher. Indeed, when poultry litter is considered, the net electricity generation can be estimated equal to 0.74 kWh/kg_{litter} or 1.13 kWh/kg_{litter}, respectively.

The capital costs of direct combustion power plants are very sensitive to plant dimension, varying from 10,500 €/kW for a small plant (100 kW) to 2,100 €/kW for a 10 MW plant. The running costs are estimated to vary from 1,600 €/kW for the former to 146 €/kW for the latter.

It should be noted that when manure is used as fuel an accurate control of emissions is required due to the large amount of NO_X generated during the process. Staged combustion is a widely used option for lowering NO_X emissions from a high-nitrogen fuel such as poultry litter. With staged combustion, combustion conditions are somewhat more reducing and less fuel bound nitrogen is converted to NO_X.

Ammonia injection under appropriate conditions also reduces NO_X emissions, and the naturally occurring ammoniacal nitrogen in poultry litter helps keep NO_X emissions low. In some cases, more rigorous NO_X control measures, such as selective catalytic reduction, may be required for poultry litter.

Gasification: in gasification, organic materials are converted into carbon monoxide and hydrogen (i.e., "syngas") that can subsequently be used for the production of a variety of fuels and chemicals. The first step in gasification is pyrolysis where volatile components are vaporized and char (fixed carbon) is produced at temperatures under 600°C. In the second step, the char and volatile products are combusted with oxygen to form CO and CO_2, generating heat for gasification. In the final step, gasification process takes place where the char reacts with CO_2 and steam to produce CO and H_2. Hydrogen and carbon monoxide are the desired product gases and they can be directly fired in a reciprocal engine or in a gas turbine for power generation [10].

A typical syngas composition produced from a plant equipped with a fixed bed gasifier having the capacity of about 4000 kg/h and using air as gasifying agent is reported in the table 2.

Table 2: Typical syngas composition.

	% volume
CO	19–22
CO_2	10–13
H_2	18–20
CH_4	0–3
N_2	50

Gasification of poultry litter at a small scale has successfully been demonstrated and the reported overall energy conversion efficiency is estimated to be 30% [10, 11]. The capital costs for a 100–500 kW plant could be estimated as 1700 €/kW, with running costs of about 180 €/kW.

4 Economic analysis

In Italy the poultry litter production of a mean holding capacity farm and that of a district can be estimated at about 300 t_{litter}/year and 95,000 t_{litter}/year, corresponding to a thermal power of about 130 kW and 38 MW respectively.

The revenue from the sale of the net produced electric power depends on the unit price assumed. In Italy the price is actually under definition and maintained between 0.08–0.15 €/kWh. The evaluation was performed considering a mean price of 0.115 €/kWh. The specific revenue referred to the unit mass of manure R_{sp} (€/kg_{manure}) can be easily calculated by

$$R_{sp}=P_{ee} \times P_{sp}$$

where P_{ee} (€/kWh) is the electric power retail price, and P_{sp} (kWh/kg_{manure}) is the specific electric power generation.

In an economic assessment, it must be considered that both direct combustion and gasification processes produce a by-product, ash, that owing to its content of phosphorus and potassium is a valuable product that can be sold as fertilizer.

The net fertilizer value of poultry litter ash at the energy plant, after accounting for transportation costs, any additional processing costs and marketing costs that may be required, will range from 19 to 60 €/t of ash. In this paper a mean value of 40 €/t is considered [12].

The results relative to the three considered energy conversion processes are reported in table 3. In the table, a small plant operating in situ for a mean holding capacity farm and an off-site district plant with a treatment capacity of 95,000 t_{litter}/year were considered. Furthermore a 10 year time frame is considered with a 5% rate of return.

Table 3: Cost analysis for the three considered energy conversion processes.

	Anaerobic digestion		Direct combustion		Gasification
Plant rating kW	50	15000	150	40000	150
Capacity factor (%)	93	98	81	95	80
Capital costs €/kWh	0.973	0.343	7.429	0.599	0.647
Annualized capital costs €/kWh	0.107	0.038	0.817	0.066	0.0712
O & M costs €/kWh	0.084	0.005	1.130	0.055	0.069
Transportation costs €/kWh	0.000	0.031	0.000	0.013	0.000
Total annual costs €/kWh/year	0.191	0.073	1.947	0.134	0.140
Ash value €/kWh	0.000	0.000	0.008	0.005	0.006
Electric power selling price €/kWh	0.115	0.115	0.115	0.115	0.115
Total	-0.076	0.041	1.824	0.014	-0.019

5 Conclusions

The preliminary cost analysis suggests that energy production from poultry litter is actually economically viable only in the case of large off-site plants the anaerobic digesters being the sole that can guarantee a return in a ten year frame. Nevertheless, owing to the increasing cost of electricity, the more and more stringent regulations about land spreading of manures and the growing of the economic value as fertilizer of the residual ash, energy production from poultry litter by direct combustion or gasification will probably appear particularly viable in the very near future.

Energy recovery (electricity and/or process steam) with ash export is a promising high-volume alternative use for poultry litter. Because of nutrient concentration and value in the ash, energy options provide a built-in mechanism for economically exporting surplus of phosphorus and associated nutrients out of concentrated poultry areas.

In the complex it could lead to an improvement of the environmental sustainability of the poultry industry in its service area.

References

[1] Steinfeld, H., Gerber, P., Wassenaar, T., Castel, V., Rosales, M. & de Haan, C., Livestock's long shadow: environmental issues and options, Report LEAD/FAO. Rome, 2006.

[2] Cabrera, M.L. & Sims, J.T., Beneficial use of poultry byproducts: challenges and opportunities. *Land Application of Agricultural, Industrial,*

and Municipal Byproducts, eds. J.F. Power & W.A. Dick, SSSA, Madison, WI, 2000.
[3] Axtell, R.C., Poultry integrated pest management: status and future. *Integrated Pest Management*, **4**, pp. 53-73, 1999.
[4] Windhorst, H.W., Changes in poultry production and trade worldwide. *World's Poultry Science Journal*, **62**, pp. 585-602, 2006.
[5] Food Outlook - November 2007. Series Title: Food Outlook (GIEWS) – 2 2007.
[6] UNA, Unione Nazionale dell'Avicoltura, 2006.
[7] Kelleher, M., Anaerobic digestion outlook for MSW streams. *BioCycle*, **48(8)**, p. 51, 2007.
[8] Jenner, M., The biotown, USA sourcebook of biomass energy, Indiana State Department of Agriculture and Reynolds, Indiana, April 2006.
[9] Barker, J.C., Methane fuel gas from livestock wastes: a summary. Publication #EBAE 071-80. North Carolina State University Cooperative Extension Service, Raleigh, NC, March 2001.
[10] Kwant, K.K. & Knoef, H., eds., *Status of Gasification in Countries participating in the IEA and GasNet Activity*, Netherlands, August 2004.
[11] Reardon, J.P., Lilley, A., Browne, K., Beard, K., Wimberly, J. & Avens J., Demonstration of a small modular biopower system using poultry litter, DOE SBIR Phase-I Final Report, February 25, 2001.
[12] Bock, B., Potential for ash value to make or break a poultry litter energy project. *Southern Bio-Products Conference*, Jackson, MS, March 22-23, 2005.

A review of the bioenergy potential of residual materials in Quebec

S. David[1] & N. Abatzoglou[2]
[1]*Laboratoires Shermont Inc., Canada*
[2]*Université de Sherbrook, Canada*

Abstract

In Quebec more than 48% of residual materials produced in 2006 throughout all societal activities have been disposed of finally in landfills. Despite of the government policies on 3Rs (Reduction, Reutilization, Recycling) and the total recovery objectives set-up, the total mass landfilled has increased by 120% over the last 10 years. Over the last 50 years, although the agricultural surface decreased by half, the number of pigs tripled. With an annual production of almost 8 million pigs, on a more than 400 out of a total of 1200 Quebec municipalities have a considerable surplus in liquid manure. The present work undertakes a review of these two growing problems-opportunities in Quebec and proposes alternative and sustainable solutions. Since the building and operation of experimental bioreactors (manure digesters) in some farms in Quebec, the anaerobic co-digestion of the rural and urban putrescible materials seems an interesting avenue. The development of a green energy vector by valorizing the so-produced biogas, would also contribute in decreasing the rejection of contaminants in the environment. In this work, only the biomass, resulting from the residential sector and the quantity of the liquid manure available in Quebec, is considered to calculate the bioenergy potential. The bioenergy output of the various methanisation processes is compared with that abroad, particularly in Europe. The available biomass in Quebec has been calculated based on available statistical data regarding the production of urban waste and manure. The review shows that this bioenergy vector technology is sustainable and that its commercialization can be profitable at a cost not higher than that of disposal in landfills.
Keywords: biomass, organic waste, anaerobic methanisation, bioreactor, codigestion, green energy, bioenergy.

1 Introduction

It is well established that actually the prices of the majority of the goods and services do not integrate the real social and environmental cost. This is ecologically unacceptable. This is mainly due to the fact that the markets, in their current structure, are unaware of the limits of the natural processes. In nature, all residual matter generated by a living organism is used as a resource by another; this ensures the natural recycling of all used resources. The challenge of our society is to reproduce this process in its economy.

The putrescible matter is one of the principal causes of contamination in the places of disposal. At the landfills, the anaerobic fermentation generates greenhouse gases (GHG). The organic compounds released by this fermentation migrate with water of leaching (leachates) and can contaminate surface as well as subsoil waters and aquifers thus, rendering them unsuitable to human consumption and to the surrounding biotopes. The recovery of the putrescible matter at ends of valorization reduces the polluting load and the GHG emissions while providing us with bioenergy and valued products (i.e. compost). The reduction of residual materials which end up in landfills is a positive element on the economy and the environment for a variety of reasons, including the reduction of the transport of these residual materials.

The hog farms activities and the resulting manure spreading for fertilizing purposes contribute considerably to the release of ammonia emissions, the increase of the phosphorous content in the soil and the pathogenic organisms in the environment. The emissions of ammonia contribute to the GHG as well as to an imbalance of the nutritive elements in the ecosystems, including surface water. Long-term liquid manure spreading in Quebec, based on directives inspired by soil nitrogen enrichment, led to an excess of phosphorus. Moreover, spreading of the liquid manure also increases the water-soluble phosphorus concentrations, which causes direct eutrophication conditions in surface waters. Currently the farm-by-farm approach stipulated in 2002 by the Law on the agricultural production led to a serious problem. Indeed, instead of calculating the rate of available phosphorus by catchment area, the quantity of phosphorus surplus is calculated based on the surface of the so-fertilized soil. This approach does not consider the type of soil, the zones easily flooded, the antibiotics present in the liquid manures and the initial quality of the surface waters (lakes and rivers) near the ground of spreading. These factors can have direct repercussions for certain rivers and more sensitive underground sheets of water. In an effort to minimize the harmful effects on the environment by the actual residual materials disposal and by the management of the liquid manure, this review tries to establish if methanisation of the organic residues and the liquid manure can be a sustainable solution.

2 Solid waste management in Quebec

In Quebec, based on data of 2006, only half of the materials which may undergo beneficiation were recovered. The landfill of the putrescible materials, the spreading of the animal manure and their storage are not sustainable solutions.

These practices contribute to higher management costs, to GHG increase and other environmental nuisances including the direct effect on the natural resources availability.

2.1 The waste recycling in Quebec

Since 1989, the year of the publication of the Integrated Solid Waste Management Policy, Quebec made a significant step in the direction of sustainable development. However, 17 years later, according to a Recyc-Quebec [1] assessment in 2006, Quebec recovers hardly 52% of its residual materials. More specifically, inasmuch as organic matter is concerned, its collection is at present under development in Quebec. This collection comprises the green residues, such as the branches, grass and sheets as well as the remainders of table. There are currently 44 composting installations in Quebec. Moreover, the cities or municipalities of Laval, Victoriaville, Lachute, Saint-Donat, Rawdon and the Island-of-the-Madeleine, have successful organic matter collection activities. However, the rate of recovery of all putrescible residual materials is as low as 8%! In 2006, the quantity of recovered organic matter rose to 0.36 million tons on a potential of approximately 4.5 million tons. It should be noted that the objectives of the Policy are expressed according to the potential which may undergo beneficiation. It is estimated that 85% to 90% of the residual materials of the various branches of activity have a potential of being valuably recovered. Recyc-Quebec announced in 2006 that nearly 13 million tons of residual materials are generated annually in Quebec. As Quebec's population in 2006 was 7.651 million inhabitants, the total rate of production of residual matters per capita and year is estimated at 1.69 tons, including industries, trade activities and institutions. A noticeable fact is that, in 2006, the rate of final disposal of these materials through landfill and incineration was estimated at 880 kg/year/inhabitant, whereas ten years earlier, this quantity was lower; 740 kg/year/habitant in 1996. Thus, the average inhabitant of Quebec, including industrial, trade, construction, restoration/demolition and institutions activities, as well as the municipalities, eliminated in 2006, 875 kg/year/inhabitant and recovered 940 kg/year/inhabitant.

2.2 The compostable material coming from the municipal residential sector

The municipal sector generates an average of 23%w/w of Quebec's residual materials. Thus, in 2006, the municipal sector generated 3.015 million tons. This represents an average of 400 kg/year/inhabitant. Although 95% of these materials are theoretically recyclable, only 31% of them were recovered. Thus, 255 kg/year/inhabitant were disposed of finally in landfills or incinerated. The putrescible materials represent 44% of the total 3.015 million tons generated annually. This is 1.327 million tons of compostable organic material per year which may undergo beneficiation. The available statistical data give us the information that only 0.110 million organic tons of materials were recovered in 2006. Table 1 presents Quebec's population by administrative areas, according to the 2001 census, and an estimation of the quantity of putrescible materials available by areas. This estimation was done by assuming that the residual material generation is geographically uniform in Quebec. Globally it is easy to calculate that the remaining available compostable material is 183 kg/year/inhabitant.

Table 1: Generation of organic compostable materials by the residential sector in the different administrative regions in Quebec.

Administrative regions	Population	Organic compostable materials (tons/year)
Bas-Saint-Laurent	200,630	36,715
Saguenay-Lac-Saint-Jean	278,279	50,925
Capitale-Nationale	638,917	116,922
Mauricie	255,268	46,714
Estrie	285,613	52,267
Montréal	1,812,723	331,728
Outaouais	315,546	57,745
Abitibi-Témiscamingue	146,097	26,736
Côte-Nord	97,766	17,891
Nord-du-Québec	38,575	7,059
Gaspésie-îles-de-la-Madeleine	96,924	17,737
Chaudière-Appalaches	383,376	70,159
Laval	343,005	62,770
Lanaudière	388,495	71,095
Laurentides	461,366	84,430
Montérégie	1,276,397	233,581
Centre-du-Québec	218,502	39,986
Total:	7,237,479	1,324,460

2.3 The landfill sites

The average maintenance costs of the landfill sites is approximately 65 cnd$/ton. The law "Règlement sur l'enfouissement et l'incinération de matières résiduelles" (c. Q-2, r.6.02) adds an additional fee (tax) of 10.22 cnd$/ton. On this basis, approximately 504 million cnd$ have been spent in Quebec only in 2006 for the landfill of the residual materials. These numbers do not include the transport cost. Moreover, these costs do not contain the management of the animal manure, mining, papermills and sawmills residues, the contaminated soils as well as all materials characterized as dangerous and biomedical. According to the "Règlement sur l'enfouissement et l'incinération de matières résiduelles" (c. Q-2, r.6.02), a soil of minimal thickness of 30 cm must cover each day the landfilled residual materials in order to protect it from vermin and biogas spreading.

2.4 The porcine industry

In June 2002, Quebec adopted the law “Règlement sur les exploitations agricoles” (REA); this law is aimed at agricultural activities within the context of the sustainable development. Its main objective is to have the farmers being more concerned about the management of the animal manure and other fertilizing matters, more particularly phosphorus. The phosphorus assessment, required by the REA, is one of the main tools of this law. The farms having a surplus of fertilizing potential will be under the obligation, according to REA, to reach a balance between the capacity of the soil to receive phosphorus and the quantities of manure to be spread out in these lands. For some of them, the only solution enabling them to reach this balance will be the treatment of the liquid manure. In Quebec, the number of farms passed from 135,000 in 1951 to 35,000 in 1996. Moreover, the agricultural surfaces decreased by about 50% since 1951; from 3.4 to 1.9 million hectares. However, during the same period the number of pigs has tripled. The annual volume of production should not be confused with the number of pigs in inventory on the porcine exploitations during the year. Quebec counted approximately 3.9 million pigs in inventory in 2004 and 4.2 million in 2006. There were 4,097 producers of pigs in 2004 and 4,309 in 2003, a fall of 212 in one year. Such a geographically uniform space reduction, in one year only, accompanied by a production increase, can be explained by the increase of the activities of the existing integrators. Indeed, since the total annual volume of animal manure appears constant in Quebec (at approximately 32 million m^3 of animal manure, of which approximately 10 million m^3 of liquid manure), the farms have to manage an increasingly larger volume of manure. The quantity of liquid manure was multiplied by a factor of 90 between 1951 and 2001! Approximately 60% of the hog farms must have to establish local agreements for manure land spreading, because they do not possess enough arable land to respect the law. According to the Quebec Ministry of the Environment, more than 400 municipalities out of 1,200 are in theoretical surplus of manure and/or liquid manure. Table 2 presents the quantity of liquid manures produced in Quebec.

In March 2006, a refundable temporary tax credit was founded for the acquisition of installations for the treatment of pig liquid manure. This credit accounts for 30% of the acquisition and the installation costs of such a system of manure treatment; a maximum of 200,000 cnd$ is allocated by agricultural establishment. Only porcine companies, which do not qualify or make use of the provisions of chapter 6 of the program Premium-Green, qualify for this tax credit. All technologies of liquid manure treatment decrease the volume of the final disposal; this means that all fertilizing elements are concentrated in smaller volumes. They also contribute to reducing the odours associated with the management of liquid manures. Denmark is a pioneer in the management of livestock wastes and the optimal use of the nutritive elements contained in them. Several successful programs have directly risen from Denmark's agricultural policy: for example, the installation of V and W shaped sludge pits leads to a reduction of the ammonia evaporation by as high as 25%, while the cooling of the sludge in the pits results in an additional reduction of 20 %.

Table 2: Manure production by region in Quebec.

Administrative regions	Number of pigs [2] (2006)	Manure (tons/year)
Bas-Saint-Laurent	153,459	475,723
Saguenay-Lac-Saint-Jean	16,954	52,557
Capitale-Nationale	97,433	302,042
Mauricie	142,945	443,130
Estrie	249,363	773,025
Montréal	n.d.	n.d.
Outaouais	365	1,132
Abitibi-Témiscamingue	n.d.	n.d.
Côte-Nord	n.d.	n.d.
Nord-du-Québec	n.d.	n.d.
Gaspésie-îles-de-la-Madeleine	n.d.	n.d.
Chaudière-Appalaches	1,296,635	4,019,569
Laval	n.d.	n.d.
Lanaudière	274,458	850,820
Laurentides	31,178	96,652
Montérégie	1,420,141	4,402,437
Centre-du-Québec	519,365	1,610,032
Total:	4,202,296	13,027,119

2.5 The methanisation industry

The management of the organic residues is not only one a concern in Quebec. In Europe there are already biogas production facilities from the anaerobic fermentation of organic residues. In Quebec, the final disposal of the majority of organic compostable material is done systematically as follows:

- Incineration for the municipal wastewater treatment sludge;
- Landfill disposal for the residential (table) residues;
- Composting for the green residues;
- Land spreading for the animal manure.

2.5.1 The legal context for the energy in Quebec

In Canada, energy is under provincial jurisdiction and the law provisions vary considerably from one province to another. The new strict regulations covering the management of waste and the high costs of energy seem to boost the interest in all provinces. In fact, the province of Ontario has recently put in application a regulation establishing the price of 0.11$/kWh (0.14$/kWh at the peak hours) for green electricity. This makes biogas more attractive to the Ontario's farmers. Ontario and Quebec are the only provinces to offer subsidies to those who install and operate biogas production systems. In fact, Quebec pays 70% of the costs incurred to install a manure handling system (maximum of 200,000$). The Ontarian producers can obtain a subsidy of up to 60,000$ for a system of manure treatment and storage capacity in the farm. Currently in Quebec, no law exists covering specifically the production of biogas. However, some regulations like the one covering the waste elimination activities (landfill and incineration) and the production of electricity contain some clauses on biogases. For example, the landfills must be appropriately sealed in order to avoid the migration of biogases through the air or the subsoil. Thus, biogases from these sites must be collected and, in the absence of energetic exploitation, this biogas must be burned safely and in accordance with the law for the atmospheric quality preservation. It is rather obvious that, without a specific regulation covering the production of biogas, and which defined clearly all constraints to be respected, it is difficult to establish a long-term biogas production system. However, according to the article 22 of Quebec's Law on the Environmental Quality (LQE), a certificate of authorization is requested for the establishment of a residues management activity. To produce such a certificate, an Environmental Impact Assessment study must be deposited to the Ministry of the Sustainable Development, the Environment and Parks (MDDEP). In the case of the construction of a high capacity power station, the project team has to deposit the preliminary draft of such a study to the Office of the Public Hearings (Bureau d'audiences publiques sur l'environnement - BAPE), such as defined in article 31.1 of the LQE. In Canada, under the article 5 of the Canadian Law on the environment, an environmental assessment is requested also for having the project authorized. If the project contributes in establishing more sustainable management activities, the Ministry of the Natural Resources Canada could even provide financial assistance to the project in order to help respecting the Environmental Law provisions.

Moreover, the Federal Government brought modifications to the rates of the deduction for goods amortization (DGA). The maximum rate of DGA for each type of redeemable good is fixed in the Law covering the field of income tax payment. The rate of DGA applicable to certain high efficiency cogeneration units and renewable energy production passes from 30% (rate of category 43.1) to 50% because of the inclusion of these goods in the new category 43.2. To qualify for this new increased rate, the material used in building such a unit must generally be new and be acquired after February 22, 2005 and before 2012. This new rate will also apply to any biogas production material and district energy distribution networks which use efficient cogeneration technology after February 22, 2005 and before 2012.

2.5.2 Biogas production

Methanisation represents a possible sustainable solution for the recycling of organic matter. All organic materials can produce biogas: i.e. animal dejections, fruits, vegetables, remainders of slaughter-house, rejections of dairy, brewery, distilling. The organic matter rich in fibre like the grass, sheets and wood is however difficult to digest and are not recommended to use as source of biogas. The biogas composition depends upon the nature of the treated waste and the treatment conditions which depend largely on the local climate. The average composition of biogas is 60% methane (CH_4) and 40% carbon dioxide (CO_2). The bioreactor is an anaerobic tank where organic matter is introduced and the temperature has to be maintained at an appropriate level depending upon the nature and the activity of the anaerobic digestion microorganisms. Under the selected conditions, anaerobic bacteria use the organic matter as feed to multiply and produce biogas as a result of their metabolism. The quantity of organic matter which exits the bioreactor is equal to that at the entrance minus the produced biogas. The digested organic matter is practically odourless, has a high fertilizing value, it is less polluting and reduced in organic load (DBO_5). The exiting solid/liquid mixture can be separated; the solid muddy part (sludge) can be composted and the liquid part can be used directly as a fertilizer or treated and disposed of safely according to the existing regulations.

The collection and energetic exploitation of the biogas produced during the anaerobic methanisation decreases the GES by a factor of just as 20. Since the digestion conditions are better controlled within a bioreactor, the total quantity of biogas per unit of organic matter extracted from a bioreactor is higher than that extracted from a landfill site. A bioreactor is 3 to 10 times more energy efficient than landfill. The quantity of biogas which can be extracted from the organic matter also depends on the type of organic residues and the performance of the bioreactor. Therefore, the production of biogas varies from 20 to 800 m^3 per ton of organic material. Each m^3 of biogas contains the equivalent of 6 kWh of energy. However the conversion of biogas into electricity by an electric generator produces 2 kWh_e; the other 4 kWh is dissipated in the form of heat and at least 75% of this which can be recovered and used for thermal energy needs. The capital investment for a bioreactor is relatively high [3]. The payback period for such an investment is estimated [4] at 7–10 years and it is due to the value of the biogas and more precisely its value as an electrical and/or thermal energy vector. Regarding the biogas as an energy vector, three uses are tested industrially: (a) combustion in a boiler for heat generation; (b) fuel of internal combustion engines for the production of electricity or for car motors and (c) in cogeneration for the simultaneous production of heat and electricity.

The treatment by methanisation of a tone of agricultural organic matter produces 500 m^3 of methane at standard conditions. The methanisation by anaerobic digestion utilizes the organic matter which is most easily digestible, that is 30–80% of the oven dry organic material. Besides, this is the part which generates organic pollution, unpleasant odours and biogas. The part remaining after methanisation (20–70%) can be used as fertilizer in agriculture. The methanisation of wastewater treatment plants sludge as well as all putrescible domestic (residential) and industrial waste reduced their quantity by up to 35%. A ton of these organic residues generates up to 175 m^3 of biogas. A landfill recovers an average of 100 m^3 of methane per ton of treated waste, whereas in agriculture, a ton of manure produces in 90 days approximately 60 m^3 of biogas containing 55% of methane.

The oldest and more mature methanisation technology through anaerobic digestion is that of "the infinitely mixed" bioreactor in whom mixing is ensured hydraulically or more frequently by biogas recirculation. This type of bioreactor is generally operated with mesophilic organisms towards 35°C. In the last years several technological novelties have been proposed and improved the efficiency of this type of bioreactor in such a way that streams like domestic organic materials, which are generally more complex than manure, could be received and methanized as well. Degrémont, for example, has proposed the upstream use of a "hydropulpor", a metal tank equipped with a central propeller which lyses the vegetable cells and puts the organic matter in full suspension in the recycle stream of the process (proceeded BTA). In this process the inorganic matters which could enter the unit (e.g. plastic, textile) float on the surface and are recovered by a hydraulic comb; the heavy mineral elements (e.g. glass, stones, bones) decant and are extracted by the bottom. The thus produced liquid pulp passes then in a hydrocyclone which removes the heavy fine and abrasive particles like sand. In Germany and Austria, bioreactors appeared in ovoid forms [5]. This form has a smaller footprint per volume, allows a more homogeneous and regular mixing thus reducing the dead zones or those of preferential fermentation and facilitates decantation. The second generation bioreactors use the technology of the fixed cells or "anaerobic filter". This technology allows a quasi-continuous flow bioreactor operation because the bacteria are fixed in a preselected support and they are not entrained by the treated liquid flow.

In France, the process suggested by Proserpol, comprises a support made up of plastic rings in bulk and the injection of the feed is carried out by the top of the bioreactor all over and through the bacterial stabilization rings. In the Netherlands, the process UASB (Upflow Anaerobic Sludge Blanket) was introduced [found in 5] in which the support is biological and flow is ascending. Compared to the old technique, this technology has the advantage of being much faster (a few hours or a few days instead of two weeks). It is possible to still speed up the process by adopting the thermophilic fermentation, which multiplies the productivity by a factor of 4–10 and produces a biogas richer in methane (up to about 80%v CH_4). For the hog farms manure, the process must proceed in two stages, the preliminary acidification, then methanisation, which call upon families of different bacteria, which must, thus, be held in distinct tanks. The Tilburg bioreactors in the Netherlands (1994, 52,000 tons/year) and that of Engelskirchen in Germany (1998, 35,000 tons/year), are fed only with organic materials coming from domestic waste. The methane produced in Tilburg's bioreactors is introduced into the gas network of the municipality for domestic use. The Engelskirchen's biogas is used for the production of electricity. Thanks to a sorting at the source, both produce a compost of high quality. During its high output operation, the factory of Tilburg produces 4 million m^3 of biogas per year. This biogas contains 56% CH_4, very small amounts of hydrogen sulphide, and 31,000 tons/year of organic soil conditioner which may undergo beneficiation in agriculture. The idea of a selective and centralized co-digestion of agricultural biomasses was born in Denmark and was applied since 1988 thanks to the support of the government. It consists in collecting in the vicinity of the manures production sites the existing domestic and/or industrial organic waste, sludge from wastewater treatment stations and methanize them together. The economic assessment of the operation is mitigated. With a starting investment of 8 million $ and operating costs of 600,000 $/an, the payback period has been 8–10 years thanks to the subsidy of 3 million $ granted by the Danish Ministry of Energy. The main part of the incomes comes from the sale of biogas (800,000 to 1,000,000 $/year) and from the waste management tipping fee which the industrialists pay to get rid of their waste (approximately 200,000 $/year).

3 Methanization in Quebec?

Methanization represents a complement of activity for the farmers, who can develop economically and energetically the liquid manure and even the organic part of the domestic residues. Methanisation is synonymous with energetic autonomy for the biogas producers. While bringing an energetic and ecological response to the problem of the treatment of the liquid manures and organic residues, methanization is an activity of depollution and an alternative to landfill.

The environmental cost of the deterioration of nature by hog farming must be taken into account in the market price of the pig. This cost can be attenuated by agreements between eco-energy the hog farms; thus, hog farms manure can be a feedstock for eco-energy farms producing compost and biogas as an energy vector. Only the cost of transport would be the responsibility of the producer, whereas the eco-energy farms would recover and market the green energy. Producers' cooperatives can also contribute financially in order to install common bigger bioreactors.

As explained below, the establishment of a “two ways” selective waste sorting at the source and pick-up could be a sustainable solution. Our residues would be split in putrescible organic and non-putrescible materials. This strategy would allow recovering nearly 95% of all residual matters in the domestic sector. The codigestion of putrescible organic residues and liquid manure would be beneficial to both the agricultural and municipal sectors. The accessibility of this putrescible resource was established by administrative regions in Quebec. The density of the population is not an important factor, since in rural as well as in urban areas, the transport of the organic matter to an eco-energy farm is at least as easy as in any existing landfill site. The only element which differentiates the administrative regions between themselves is that certain among them do not have or have few hog farms, which would be favourable to install and operate bioreactors because of their characteristics as large urban number of eco-energy farms is required per administrative region and potential revenues. However, due to the proximity of the territories of Montérégie and Lanaudière, the administrative areas of Montreal and Laval, which are such urban centers, could easily transport their organic residues in bioreactors operated in these two other areas.

Table 3 presents the energetic potential of the combined domestic putrescible organic material and the manure in Quebec. Each ton of organic matter was estimated as having a potential of production of 100 m^3 of biogas. This is very conservative because it represents the equivalent of the production in a landfill which is at least 3 times lower than that in a bioreactor. The energy which can be recovered by biogas, was evaluated to 2 kWh/m^3. Thus, according to table 3, an energy potential of 2,844,656,000 kWh is available in Quebec. Considering that in Quebec, a one-family residence consumes 17,000 kWh annually, the energy needs of about 167,000 houses can be covered. Since the cost of residential electricity in Quebec is approximately 0.05 cnd$/kWh, the green energy worth is roughly 142 million of cnd$.

Table 3: Energetic potential of the domestic putrescible organic material and the manure in Quebec.

Administrative regions	(a) Domestic organic residues (tons/yr)	(b) Revenue Residential Sector (M$cnd/yr)	(c) Manure (tons/yr)	(d) No	(e) Capital cost M$cnd)	(f) Power (MWh)	(g) Revenue from energy (M$cnd/yr)	(h) Mixed Revenue (M$cnd/farm/yr)
Montérégie (+Montréal)	565,309	36.7	4,402,437	99	29.7	993,549	49.7	0.573
Chaudière-Appalaches	70,159	4.6	4,019,569	82	24.6	817,946	40.9	0.255
Centre-du-Québec	39,986	2.6	1,610,032	33	9.9	330,004	16.5	0.279
Lanaudière (+ Laval)	133,865	8.7	850,820	20	6.0	196,937	9.8	0.625
Estrie	52,267	3.4	773,025	17	5.1	165,058	8.3	0.388
Bas-Saint-Laurent	36,715	2.4	475,723	10	3.0	102,488	5.1	0.450
Mauricie	46,714	3.0	443,130	10	3.0	97,969	4.9	0.490
Capitale-Nationale	116,922	7.6	302,042	8	2.4	83,793	4.2	1.175
Laurentides	84,430	5.5	96,652	4	2.1	36,216	1.8	1.300
Saguenay/Lac-Saint Jean	50,925	3.3	52,557	2	0.6	20,696	1.0	1.850
Total :	1,197,292	77.8	13,025,987	285	86.4	2,844,656	142.2	0.469

(a) Tons of domestic organic materials = Table 1
(b) Revenue residential sector = 65$cnd * tons of domestic organic materials
(c) Tons of manure = Table 2
(d) Number of farms = (tons of domestic organic materials + tons of manure)/50 000 tons
(e) Capital cost for bioreactor installation and operation = 0,300 M$cnd * number of eco-energy farms
(f) Power = (tons of domestic organic materials + tons of manure) * 100 m^3 of biogas * 2 kWh
(g) Revenue from energy = Power * 0.05 $cnd/kWh
(h) Mixed Revenue = Revenue from energy + Revenue from domestic sector – Capital cost for bioreactor installation and operation

The number of eco-energy farms was evaluated according to a self-capacity of 50,000 tons of organic matter per year. The cost for the establishment of such a farm was estimated at 0.3 million cnd$. It should be noted that the actual cost of the management of the organic matter coming from the domestic (residential) sector is equal to 90% of the capital cost of the installation of the 285 bioreactors necessary for methanizing annually the 14 million tons of the available putrescible materials (domestic + manure). Considering the contribution of income of the residential sector, only the areas of the Boiler-Appalachian Mountains and Center-of-Quebec, do not offer an income higher than the cost of establishment of an eco-energy farm.

4 Conclusion

The establishment of a two ways (putrescible and non-putrescible) selective sorting and pick-up of the domestic residues in Quebec would increase the rate of recovery of all the residual matters. The non-putrescible matters can be recycled up to 95%, whereas the organic matter can be used as a renewable energy vector. The only problem comes from the green residues (grass, sheets, paper and paperboard) which contain a high percentage of non-easily digestible collection which is already established in some municipalities (i.e. Victoriaville, Sherbrooke). A good sorting of the putrescibles at the source (residential) is important because the efficiency of the methanization bioreactors and the quality of the resulting compost are strong function of the feedstock quality. The existence of incentives and the appropriate enforcement of a regulation establishing an obligation to treat the manure of all hog farms in eco-energy farms using bioreactors would also contribute to the reduction of contaminants in the environment. All putrescible matters (domestic putrescible organic materials and liquid manure) can be treated (methanized) in eco-energy farms, equipped co-digestion bioreactors. As it is well established that the ideal operation conditions of the anaerobic digestion of these two distinct families of putrescible organic materials are not the same, some additional fundamental and applied research is necessary to have a sustainable operation. The existence of two parallel bioreactors might be an appropriate solution since it is known that different bacterial populations are needed. By considering only the income received from the production of green energy, the payback period of a bioreactor at an average Quebec hog farm is approximately 10 years. As for the eco-energy farms, they could require higher initial investments because of the combined feed complexity. These additional costs can be set-off by the perception of the equivalent landfill disposal tipping fee which in Quebec is in average 65$/ton.

If this scenario becomes a reality, nearly 13 million tons of liquid manure and 1 million tons of putrescible domestic organic materials produced annually in Quebec could become a source of green, renewable energy in a sustainable development context, without any economic or social sacrifice.

Acknowledgements

The contribution of Steve Boivin, Eng. and Richard Royer, Eng. of Bio-Terre Systems Inc., (150, rue Vimy, Sherbrooke, Québec, Canada, J1J 3M7) is gratefully acknowledged.

References

[1] Recyc-Québec, *Bilan 2006 de la gestion des matières résiduelles au Québec*, Bibliothèque et Archives nationales du Québec, 2007.
[2] Institut de la statistique du Québec, *Inventaire de fin de semestre de porcs, par région administrative, Québec 2005-2006*, Gouvernement du Québec, 2007.
[3] Électrigaz, *Le biogaz est une énergie renouvelable issue de la décomposition naturelle de la matière organique par des bactéries anaérobiques.* 2007. http://www.electrigaz.com
[4] Électrigaz, *Le biogaz est une énergie renouvelable issue de la décomposition naturelle de la matière organique par des bactéries anaérobiques.* 2007. http://www.electrigaz.com
[5] Énergie Plus, *Rapport, études France,* 2007. http://www.energie-plus.com

Waste as a resource in the United Kingdom: energy and sustainability

A. Arias-Garcia & R. Lambourne
Waste Management Department, RPS Group PLC, UK

Abstract

In the world in which we live, where environmental issues are becoming increasingly important, seeing waste as a resource and recovering as much of its value as possible is an imperative. The aim of this paper is to offer a review of the waste management activities currently being undertaken in the United Kingdom (UK) and to provide a viewpoint on the reasons for technology choice. These reasons should be a key study focus to provide the tools for refining and improving practices. From an environmental viewpoint and looking at the global warming potential of any waste management solution it is clear that the choice should always aim to minimise greenhouse gas (GHG) emissions to the atmosphere. However, facilities are still being built where the choice is far from the best environmental solution. Among the factors affecting the technology choice, an important one is legislation, though it can still serve as a barrier to the successful integration of different mechanisms; in other cases the answer comes from technology limitations or economic balance. Analysing the outcomes and assessing the reasons for the technology choice in the UK is an essential exercise if progress towards a truly sustainable pattern of waste management is desirable. The authors have identified a number of issues that affect waste management in the UK namely the implementation of the Landfill Directive, the Waste Incineration Directive, renewable energy incentives and the unproven status of some technologies. The UK market represents a great opportunity for European technology providers and other sector stakeholders. However, understanding the UK business model is essential. This analysis could help European businesses that are achieving excellence in other countries to find the appropriate pathways to successfully bring the best solutions to the UK.
Keywords: waste, energy, renewable, sustainability, UK, policy, efficiency.

1 Waste arising and management in the UK

Sustainable waste management is high on the political agenda and includes issues of sustainability, waste growth and public concern over proximity to proposed waste treatment facilities as well as GHG emissions.

The UK produces around 330 million tonnes of waste annually – a quarter of which is from households and business. The rest comes from commercial and industrial, construction and demolition, sewage sludge, farm waste, spoils from mines and dredging of rivers (fig. 1). The UK is making progress towards the goals of the waste hierarchy and there is a steady trend towards reducing landfill and increasing recycling.

In 2006/7 local authorities (LAs) in England collected 29.1 million tonnes of waste from household collections and civic amenity sites, according to estimates released by the Department of Food and Rural Affairs (Defra). Of that, household waste accounted for 25.9 million tonnes (89%). The proportion, which was recycled or composted, rose from 27.1% in 2005/6 to 30.7% in 2006/7. The percentage, which ended up in landfill, fell from 62% (17.9 million tonnes) in 2005/6 to 58% (16.9 million tonnes) in 2006/7.

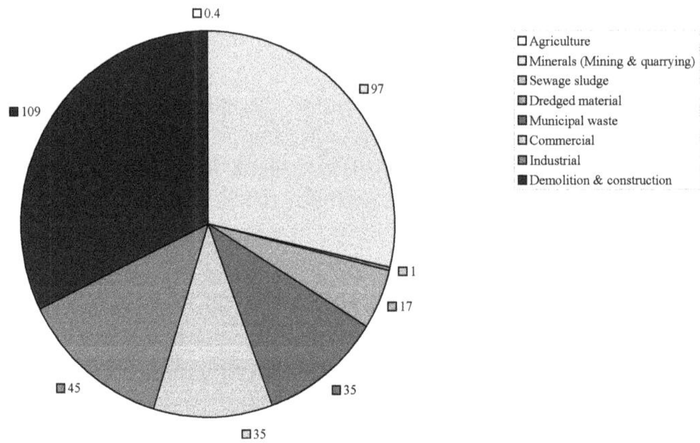

Figure 1: Estimated total UK annual waste arisings (million tonnes) by sector 2002/2003. Source: Derived from data published by Defra [2].

LAs control the primary supply of municipal solid waste (MSW), often under a two-tier system. In England waste disposal or treatment contracts are let by the Waste Disposal Authorities (WDAs), which are mainly the County Councils, with the District Councils being responsible for collection including kerbside recycling. Unitary Authorities and Metropolitan Districts are responsible for collection and disposal as are the local authorities in Wales and Scotland.

The last waste survey encompassing commercial and industrial (C&I) waste was in 2002–2003. C&I waste in England accounted for 68 million tonnes (fig. 2) with 38 million attributable to industry (of which 13 million tonnes was landfilled) and 30 million to commerce (of which over 14 million tonnes went to landfill).

C&I waste can include paper, card, metals and scrap equipment, mineral wastes and residues. It is produced by the manufacturing industry, wholesale and retail, catering, educational and administrative sectors.

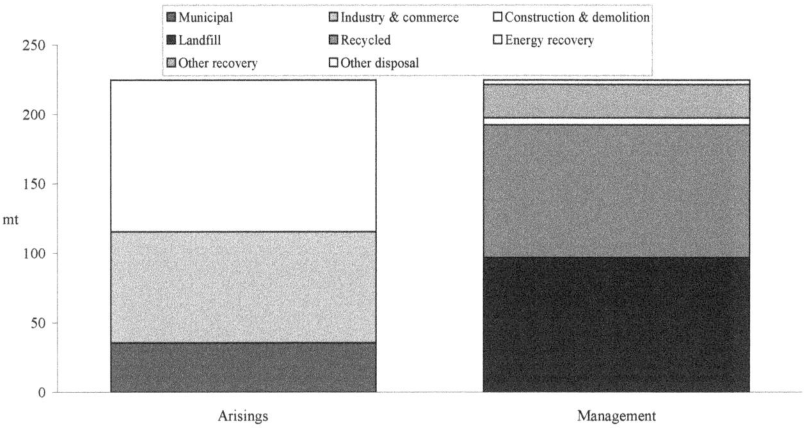

Figure 2: Arisings and management of waste from municipal and commercial and industrial sources 2002/3 Source: Adapted from data published by Defra [2].

Looking at fig. 3 and fig. 4, it can be seen that the individual sector that produced the most waste was the retail sector (commercial), which generated 13 million tonnes followed by food drink and tobacco (industrial), which generated 7 million tonnes.

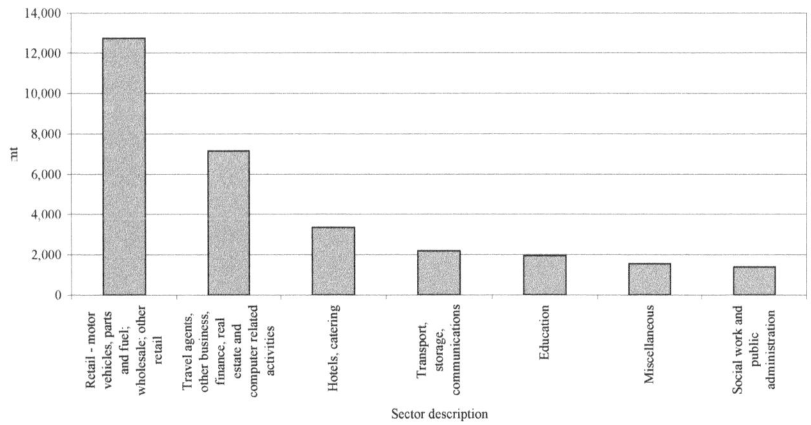

Figure 3: Commercial waste arisings by business sector. Source: Derived from data published by Defra [2].

Commercial sector waste contracts benefit from straightforward negotiations in contrast to the lengthy local authority procurement process. However, depending on the nature of the commercial waste, it may be that treatment business will still have to enter into a contract with a fuel producer and/or local authorities.

The key issue in sourcing C&I waste will be the length of contracts available. Any facility to process C&I waste arisings will probably have to contract on the basis of short term deals or even the spot market where price is very competitive.

To date, as there have not been any target requirements for C&I waste, virtually all material ends up in landfill as this has represented the cheapest option. However with the increasing price of landfill and Landfill Tax, it is considered likely this source of material will become increasingly attractive.

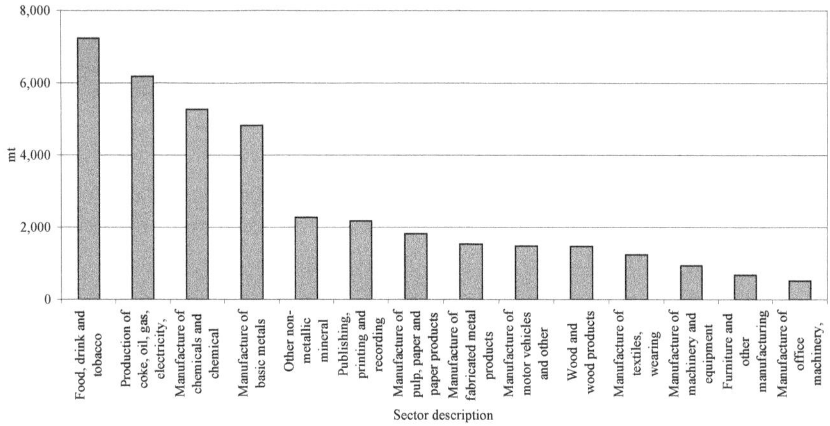

Figure 4: Industrial waste arisings by business sector. Source: Derived from data published by Defra [2].

2 Energy usage in the UK

In the UK the primary demand of energy in 2006 was 244.0 million tonnes of oil (10 million TJ and 3 million GWh). Total final consumption in 2006, which includes non-energy use of fuels, was 169.6 million tonnes of oil (7 million TJ and 2 million GWh). Final energy consumption in 2006 was mainly accounted for by the transport sector (35%), the domestic sector (27%), the industrial sector (19%), the commercial sector (6%) and non-energy use (7%) [2].

The main fuels used by final consumers in 2006 were petroleum products (48.5%), natural gas (31.5%) and electricity (17.5%). Of the petroleum products consumed by final users, 13.5% was for non-energy purposes; for natural gas 1.5% was consumed for non-energy purposes.

In 2006, the main industrial groups consumed 32.6 million tonnes of oil, (1 million TJ or 0.4 million GWh). The largest consuming groups were chemicals (18%), metal products, machinery and equipment (12.5%), food, beverages and tobacco (11.5%), iron and steel and non-ferrous metals (9%), mineral products (8%), and paper, printing and publishing (7%). The figures are illustrated in fig. 5.

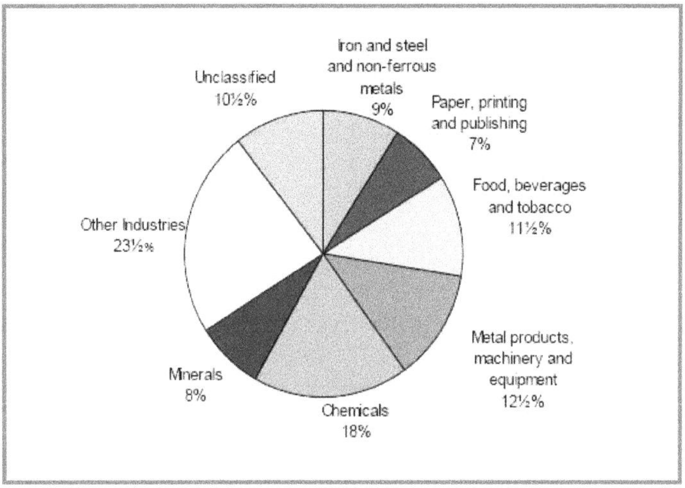

Figure 5: Energy consumption by main industrial groups 2006. Source: Department for Business Enterprise and Regulatory Reform [2].

3 So why is it so important to relate both?

In March 2008 the Cabinet Office of the United Kingdom published the report "The National Security Strategy of the United Kingdom" in which climate change and competition for energy are among the factors discussed. The report describes climate change as potentially the greatest challenge to global stability and security. In terms of energy, it states that on present projections global energy demand will be more than 50% higher in 2030 than today [1].

UK statistics show that in 2006, total carbon dioxide (CO_2) emissions were 554.5 million tonnes. 90% of this total was accounted for by three sectors, business, transport and residential, which represented 35%, 28% and 27% of the total, respectively. Of these sectors, there was an increase of 1.4% in business emissions, from 193.4 million tonnes in 2005 to 196.1 million tonnes in 2006, but little change in residential or transport emissions.

Total methane (CH_4) emissions in 2006 were 2339.8 thousand tonnes. 41% of these emissions were from waste management and 38% from agriculture. Of these major contributing sectors, there was a 1.4% increase in agricultural emissions, from 878 thousand tonnes in 2005 to 890 thousand tonnes in 2006, but little change in emissions from waste management.

Emissions of nitrous oxide (N_2O) were 123.5 thousand tonnes in 2006, a 3.7% decrease from 2005 levels. Emissions from agriculture, which accounted for 68%

of total emissions on an end-user basis in 2006, fell by 4.6% between 2005 and 2006. Emissions from the other major contributing sector, transport, were 19.0 thousand tonnes in 2006 (15% of total emissions), a 2.0% increase on 2005 levels.

Fig. 6 shows the UK emissions of CO_2, CH_4 and N_2O by source. The plot has mixed units so all emissions can be appreciated. It can be seen that waste management and energy can play an important part in reducing GHG.

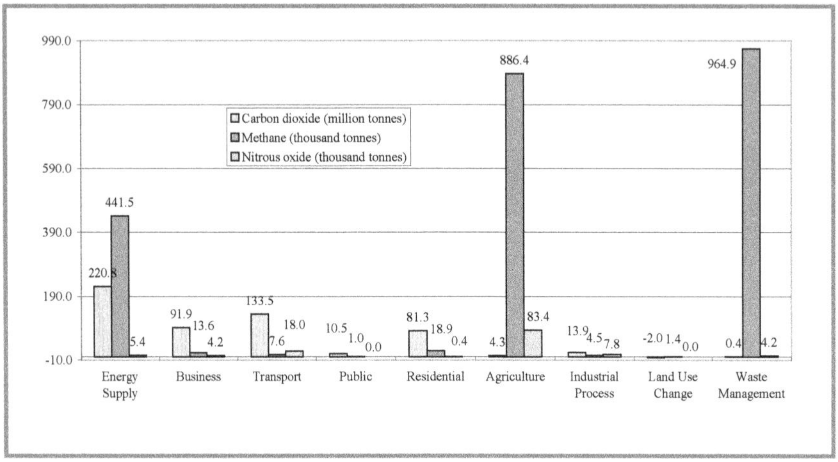

Figure 6: UK emissions of carbon dioxide, methane and nitrous oxide by source. Source: Derived from data published by Defra [6].

Fig. 6 clearly shows that reducing the amount of waste going to landfill could contribute significantly towards the reduction of GHG emissions. Indeed disposal to landfill is the last option in the waste hierarchy.

The waste hierarchy provides the basic steering framework of waste policy. The existence of a clear priority of reuse and recycling over the use of waste as an energy source is essential. Reuse and recycling avoids the use of energy and new materials.

But even after reuse and recycling there is a residual that needs to be dealt with. The Defra waste hierarchy (fig. 7) includes composting at the same level as recycling. Composting will deal with some of the organic fraction of the residual. However, a significant fraction of material organic and non-organic will remain that will require treatment.

Energy-from-waste (EfW) refers to any waste treatment that creates energy from a waste source that would have otherwise been landfilled. Some EfW processes result in usable fuel products, such as methane, methanol, ethanol or synthetic fuels upon completion of the process.

Generally energy can be recovered from waste by thermal (combustion, gasification or pyrolysis) or biological treatments (anaerobic digestion). There is also a type of Mechanical Biological Treatment (MBT) plant where the output can be used as a fuel and therefore recover energy from it. This fuel can be called Solid Recovered Fuels (SRF) or Refuse Derived Fuels (RDF) depending on whether it is made to European Standards or it is simply fuel derived from nonhazardous waste that has had some form of processing. The market for SRF is not yet established in the UK but it is growing. It is a material that is of interest to a number of industries due to its quality as a fuel combined with its high biomass content.

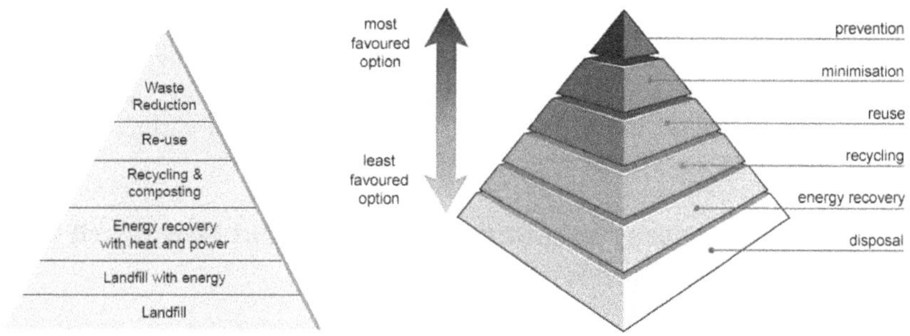

Figure 7: Waste hierarchy from Defra and Waste Framework directive. Source: Defra and Waste framework directive.

Environmental organisations like Friends of the Earth focus on phasing-out residual waste and agree with generating energy from waste via biological treatment but not via thermal treatment (on the basis, among others, that combustion creates GHG and other pollutants that cause health problems). They believe that the UK needs small-scale modular technologies to maximise removal of recyclables and to remove the biological activity of what remains prior to use in low-grade land treatment or disposal to landfill.

It cannot be disputed that thermal treatment creates GHG, as in principle, combustion is the conversion of hydrocarbons into CO_2 and water. However one would expect the Waste Incineration Directive to be robust enough to prevent any waste combustion facility from causing health problems – and if this is not the case, surely the Directive should be made stronger. On the other hand, if the energy recovered from waste is replacing energy obtained from fossil fuels why put the waste in the ground?

In terms of electricity generation, conventional EfW plants i.e. incinerators that produce electricity-only are inefficient plants – their efficiency rate of 24% compares with 38% from a modern coal-fired power station and 50% from a modern gas-fired power station. However, in Europe plants tend to use also the heat produced. The thermal utilisation is therefore typically 50–60%. In addition there are examples of innovative process design like the Zabalgarbi plant in Bilbao where the efficiency achieved is 44%. Of course there are barriers (which will be discussed later in the paper) to the implementation of these types of facilities as well as the use of heat, but perhaps government policy makers should re-examine the barriers and propose mechanisms for overcoming them.

In the UK almost half of all CO_2 emissions arise from the use of heat. That is energy used for space and water heating, industrial process heating, industrial drying and similar purposes. Therefore, if long-term reductions in carbon emissions are to be delivered, how heat is produced has to be looked at.

Power stations are burning a material that has been in the ground for millions of years and is, in human timescales, non-renewable. Assuming electricity is generated from waste with the same or higher efficiency than from coal/gas as part of an integrated waste management policy where completion of the hierarchy's initial steps are ensured, what would be the issue with EfW?

The discussion so far leads to a simple conclusion. The mitigation of GHG emissions from waste and its contribution towards reducing GHG emissions in general must be addressed in the context of integrated and sustainable waste management practice supported by a strong policy framework.

4 How much energy is contained in waste and how much is the UK already recovering?

According to the figures presented in this paper 16.9 million tonnes of household waste was landfilled in England in 2006/2007. The figures for C&I waste are from 2002/2003 (but there is no indication that the amounts have reduced significantly) with 13 million tonnes of industrial and 14 million tonnes of commercial landfilled at the time. Table 1 presents the energy content of some waste streams.

Table 1: Energy in waste. Source: EC RDF, current practice and perspectives.

Material	CV (MJ/Kg)	Material	CV (MJ/Kg)	Material	CV (MJ/Kg)
Tyres	28.5–35	Animal waste	16–17	Commercial waste	16–20
Plastics	29–40	Source separated waste	20–23	RDF from industrial waste	18–21
Waste wood	15–17 (10–15% residual water)	RDF from MSW	13	Demolition waste	14–15
Paper	12.5–22	Untreated MSW	8–11	Waste Oil	35
Paper sludge	8.5	SRF	5–25	Solvent	27
Dried sewage sludge	16–17	Household waste	12–16		

If simple calculations are applied to the figures, it could be easily stated that, out of the 16.9 million tonnes of household waste, if 25% of it would be treated via thermal treatment with an average efficiency of 50%, 4690 GWh would be obtained. Performing the same calculations to C&I waste, it could be seen that 9375 GWh could be recovered making a combined 14069 GWh which represents approximately 4% of all the industrial energy consumption.

Table 2 provides a summary of the combined capacity and energy produced from waste in the UK. Fig. 8 provides a map with a geographical representation of the waste treatment facilities in the UK.

Table 2: Summary of waste treatment facilities in the UK. Source: RPS.

Technology	Number of Facilities	Combined Capacity	Electricity produced (MWh)	Heat recovered (MWh)	Average efficiency %
Incineration	22	4403	309	90	35
ATT	6	363	50	1.5	39
AD	6	233	6	3	NA

The data presented highlight the need for more treatment facilities in the UK. There are a number of LAs that are in the procurement process and others which have just announced their preferred bidder. In addition, there is an important opportunity deal with C&I waste in a sustainable way.

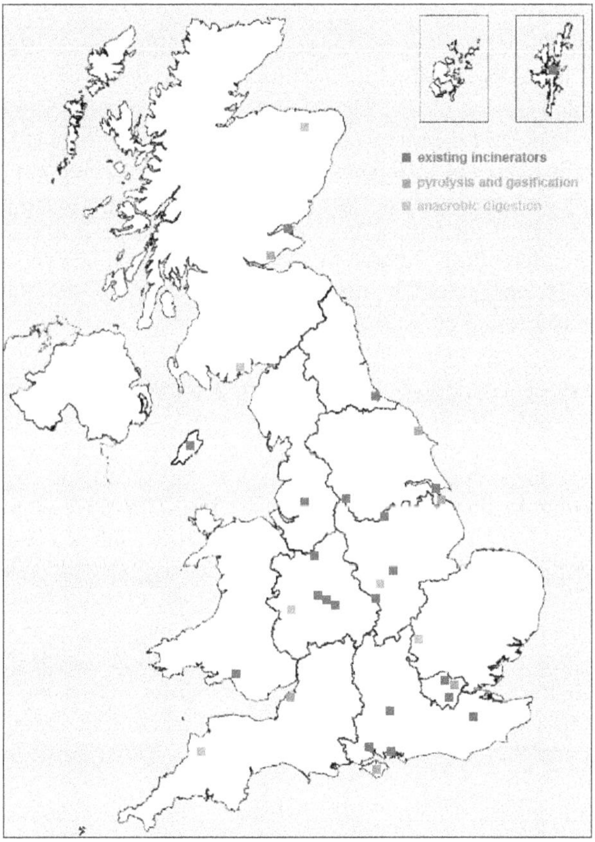

Figure 8: Waste treatment facilities in the UK. Source: RPS.

Growing numbers of UK local authorities are now selecting EfW as the most cost-effective and practical solution for treating residual waste. However some of the new proposals do not incorporate heat recovery as an option or consider efficiency as the first priority.

Drivers are different for MSW projects compared to C&I projects. In terms of MSW projects, the main driver is the diversion of biodegradable material to landfill followed by economic outcome. However, when it comes to C&I waste, it is the economic outcome that prevails. Environmental impact also plays a part in the procurement/contractual process but generally does not exert enough influence to determine the technology of choice. If sustainability was the first priority, in an ideal society, it might be able to achieve the Friends of the Earth aim of phasing out residual waste. However, in the present society with sustainability as the first priority, work could be done towards an integrated sustainable solution where waste management, energy generation and combating climate change would be jointly assessed.

5 Barriers to increasing efficiency

There are a number of barriers to this goal and the following paragraphs discuss some of them. The order of presentation implies no hierarchy of importance.

The first one is economic viability. For a project to be successful it has to be economically viable to secure the support of the financial institutions. Unfortunately there are not sufficient economic incentives to support facilities that are more costly and could achieve above-average levels of efficiency. This point links directly with legislation. As seen in table 3, there are a number of mechanisms that have been created to encourage sustainability in the waste and energy sector. However the UK has to resolve an anomaly that exists for those supplying renewable energy. Currently, a business providing renewable electricity will be rewarded through the Renewables Obligation Certificates (ROCs) as long as the electricity can feed the National Grid. Biogas and biomethane do not enjoy the same privileges and the department for Business Enterprise and Regulatory Reform (BERR) is currently studying potential mechanisms to support the development of renewable heat.

And finally project risk both to secure the material and to secure the energy user. The barriers seem to be the lack of strategic partnership at a regional level to bring together the material and reduce the risk for the operator and the finance provider. In terms of heat the threat lies in securing the user throughout the life of the project. This risk has to be minimised if heat is ever going to be utilised.

There are two recent mechanisms that might help reducing this risk by encouraging/helping energy users and waste facilities to include heat in their search/production of renewable energy. The first one is the Carbon Reduction Commitment and the second is the establishment of joint waste authorities in England.

Announced in the Energy White Paper 2007, the Carbon Reduction Commitment Scheme (CRC) applies mandatory emissions trading to cut carbon emissions from large commercial and public sector organisations. It covers around

10% of the UK economy-wide emissions and provides incentives for organisations to increase energy efficiency. The scheme started in January 2010 with a three-year introductory phase featuring simple fixed-price sales of allowances. From 2013, there is a Government imposed cap on the number of allowances and all allowances will be sold each year via an auction, a world-first for this type of scheme [3].

Table 3: Policy instruments in the UK and European Union (EU). Source: RPS.

	EU	UK
Waste	Landfill directive Thematic Strategy Waste Framework Directive Soil Directive CEN 343 Standards Waste Incineration Directive	Waste Strategy for England 2007 UK Biomass Strategy Landfill Tax
Energy	Co-generation Directive Biomass Directive Biofuels Directive Renewable Energy Directive Proposal for a Directive on the promotion of the use of energy from renewable sources	Energy White Paper Renewable Obligation Reform of the Renewables Obligation Heat Obligation
Environment	EUETS Green Paper: Adapting to climate change in Europe – options for EU action Climate action and renewable energy package	Climate Change Programme Carbon Emissions Reduction Target Carbon Reduction Commitment Climate Change Bill Planning Policy Statement. Planning and Climate change

On the other hand, in March 2008, the UK government launched a consultation regarding proposals for the establishment of joint waste authorities in England. Joint working on waste is becoming increasingly important in the drive for more efficient and sustainable waste management. While there is already a wide range of active waste partnerships across the country, one of the key barriers to joint working is the limited legal basis for joint working arrangements. This can create difficulties when authorities are dealing with the private sector. Allowing authorities to enter into a legal commitment to work together on a statutory basis provides greater certainty to potential finance providers and contractors [4].

6 Additional factors influencing the change

Historically landfill was relatively cheap and benefited from low processing costs. However, landfill tax has increased significantly in recent years (£24 per tonne in 2007/08) and will increase at £8 per tonne per year from 2008/9 to 2010/11. Furthermore, there is a shortage of landfill capacity in some parts of the UK, which means that landfill gate fees are also increasing.

Another major influence is the fact that a consistently high world oil price will affect the economics of EfW to such an extent as to overcome the economic risk. As the price of oil reaches new highs, one can see EfW projects extracting every single MWh from the energy in the waste and overcoming the current economic barriers associated with combined heat and power installations and other innovative projects.

Finally, the time will come when the mechanisms to support renewable energy will include heat. This would be coupled with other mechanisms where heat users will be encouraged (through carbon policies) to make use of renewable heat instead of producing new.

Legislation is moving fast and the UK and EU are making every effort to integrate policies and provide the mechanisms towards sustainable waste management. The UK is a market with plenty of opportunities, which is moving quickly towards sustainable waste management. The decisions made today need to be forward-thinking enough to take advantage of the change in drivers and the future integration of policies.

References

[1] Cabinet Office, *The National Security Strategy of the United Kingdom*, Crown, 2008.
[2] Department for Business Enterprise and Regulatory Reform, *Digest of United Kingdom Energy Statistics 2007*, National Statistics, 2007.
[3] Department for Food and Rural Affairs, *Action in the UK – Carbon Reduction Commitment*, http://www.defra.gov.uk, 2008.
[4] Department for Food and Rural Affairs, Consultation on draft Regulations and draft guidance for proposals for establishing Joint Waste Authorities in England, http://www.defra.gov.uk/corporate, 2008.
[5] Department for Food and Rural Affairs, *e-Digest Statistics about: Waste and Recycling*, http://www.defra.gov.uk/environment/statistics, 2008.
[6] Department for Food and Rural Affairs, *e-Digest Statistics about: Climate Change*, http://www.defra.gov.uk/Environment/statistics, 2008.

Perspectives of refuse-derived fuel in Romania after entrance into the EU

E. C. Rada[1], M. Ragazzi[1] & I. A. Istrate[2]
[1]*University of Trento, Italy*
[2]*Politehnica University of Bucharest, Romania*

Abstract

On 1 January 2007, Romania entered the European Union. For that reason Romania must comply with all the EU directives in the field of the management of the environment. In 2007, in Romania, Municipal Solid Waste (MSW) was generally landfilled without pre-treatment and energy valorization absent. Some initiatives are in progress but landfilling is still dominant. The entrance in the EU gave an important push to the industrial sectors, increasing the demand of fuels. In this frame an important option regards the use of Solid Recovered Fuel (SRF) as industrial fuel substitute. In Romania, SRF could be generated from MSW using the bio-drying process with post-separation of inert fractions as the percentage of organic fraction in the residual MSW is significant. Today in Romania bio-drying plants for residual municipal waste (RMSW) are under discussion but not yet implemented. The present paper deals with the potential use of SRF in Romania. Some aspects of local and global environmental impact related to a partial substitution of conventional fuels with SRF for industrial uses are discussed too. Some considerations are also based on the assessment of the Romanian SRF characteristics that could be potentially generated. The scenarios that are discussed concern the use of SRF obtainable taking into account also the development of the Romanian MSW composition and the effects of the selective collection evolution.
Keywords: bio-drying, cement factory, LHV, Romania, RMSW, SRF.

1 Introduction

On 1 January 2007, Romania became one of the European Union (EU) members. According to the European Association Agreement, the Romanian development policies must be guided by the principle of sustainable development, taking full account of environmental considerations [1].

Romania started to implement the EU principles on waste management before entering in the EU. In 2004, the first plan for waste management for the years 2003–2013 was issued through the national regulation 1470/2004, taking into account the national environmental protection and the sustainable development. In the first instance, waste prevention was indicated, in the second, recycling and energy generation and only in the last one, disposal of waste with no recovery of either materials and/or energy. In 2004, through the regulation 170/2004, the rubber management was imposed and in 2005 the pathway for the adoption of the directive 2000/76/EU on the incineration and co-incineration of waste was completed through regulation 268/2005.

Between 2005 and 2006, the regional waste management plans were developed followed by the county waste management plans in 2007–2009 and by feasibility waste integrated management studies to be financed through "Sectoral Operational Program Environment – SOP ENV 2007–2013". The SOP ENV helps Romania to achieve the EU objectives regarding the protection and improvement of environment quality and to develop basic infrastructure to European standards.

The last regulation issued in Romania, n. 870/2013, regarding the waste management for the period 2014–2020, imposes the up-date of the national waste management plans with the aim to reduce conventional resource consumption and to apply strictly the waste hierarchy. Until 2015 Romania has to develop 30 integrated systems for the waste management at regional and county level. Romania has already obtained a transition period until 2017 for the waste landfilling through the regulation 349/2005 and until 2015 for the import, export and waste transport through regulation 259/93. In the regulation 870/2013 it was indicated that Romania could replace most of the thermal energy produced by conventional plants with the one from waste thermal treatment plants that could treat about 15%–20 % of generated waste. In this way the request from the EU legislation 2009/28/EC regarding the use of renewable energy and the ones from the Revised Waste Framework Directive regarding the valorisation of household waste, as a percentage of the total household waste equal to 50% by 2020, could be complied.

The mechanical biological treatments (MBT) represent the most used options for the production of "fuel" starting from municipal solid waste (MSW). To this concern a one stream option of MBT could be preferred to a two stream option of MBT as the first one could put to zero the need of landfill volumes [2]. A typical one-stream option is based on the bio-drying process [3–5]. The final material from these processes, named generally solid recovered fuel (SRF), after or without a refining stage (depending on the input materials) can be sent to different thermal plants for energy recovery in concordance with the UNI norms [6–10].

Since 2006, a part of the Romanian industry has begun to support and apply an environment policy putting on the first place the reduction of atmospheric pollution, the ecological products production and the co-combustion of waste for saving natural resources. An important step that the Romanian industrial sector made was the one of a good partnership with the waste sector. The first kinds of waste authorized to be used in the industrial sector as alternative fuel are exhaust oil, sludge, used tires, refuses of the wood sector, paper and also some waste from the chemical industry. Generally the use of these fuels is limited to the cement sector. Good examples were given by the Romanian cement factories as:

- Carpatcement Holding cement factory modified its production process in 2004 in order to use also alternative fuels as rubber, plastic, textiles, paper and cardboard, wood, etc. [11];
- Holcim cement factory, has obtained the ISO 14001 certification before the aderation of the Romania to the EU, and today uses about 100 types of waste in its production processes [12];
- Lafarge cement factory implemented different standards for waste use in its plant and also developed a new company, ECOGEST, for the waste valorization [13].

Selective collection (SC) in Romania has a very low efficiency, and for that reason the residual MSW (RMSW) has a high organic fraction content: about 50% in urban areas [14, 15] because of the very low food waste selective collection. For this reason the bio-drying process could be developed successfully as demonstrated by different literature researches [5, 16–20]. After a refining stage SRF can be sent to coal fired power stations or to cement kilns for partially substituting coal and pet-coke [7, 8].

The SRF valorisation can be viewed for Romania as a strategic component of an integrated waste management policy because in this way the quantity of the biodegradable materials that could arrive in a landfill can be reduced as requested from the Landfill Directive from 1999/31/EC, the produced energy can be from renewable sources (the biodegradable part of SRF is considered as biomass) and lastly but most importantly, natural resources use can be decreased.

In this paper an option is presented and discussed for using the Romanian SRF that could be obtained from the RMSW bio-drying.

2 RMSW bio-drying and SRF

The situation of Romanian MSW from 2006 to 2011 is presented in fig. 1. Also today a part of the produced MSW is still uncollected (about 10%) [21]. The collected waste in 2011 was disposed of in 31 sanitary landfills and in 106 not controlled ones (75 of these were closed in 2012 and the remaining will be closed until 2017). During the closing of the old uncontrolled landfills, some transfer and sorting stations were constructed: 70 are operating since 2012 [21].

Thanks to the Phare CES 2004 Programme and integrated waste management projects at county level (financed by ISPA or SOP Environment) in 2012, 70 platforms operated for composting biodegradable MSW and about 50 plants and composting platforms were in various stages of implementation and regulation in terms of environmental protection [21]. However all these plants treat only 10% of the produced biodegradable waste. For this reason, the bio-drying treatment with SRF production and its energy valorisation seems to be the most viable solution for the Romanian MSW [18, 19, 22].

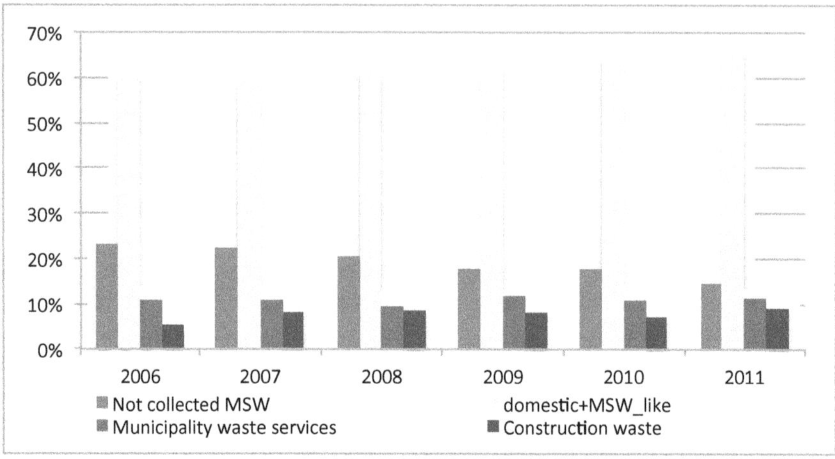

Figure 1: MSW distribution in Romania in 2011.

Romanian RMSW has generally high organic fraction content (about 50%). Fluctuations can depend on selective collection, but this sector is not yet developed to levels that can significantly affect the RMSW composition as in other EU countries [23–25]. In order to study the viability of bio-drying for generating SRF for Romanian RMSW, this process was applied at pilot scale [1, 3, 5, 18, 19].

The composition used as reference for the Romanian RMSW was calculated starting from the Romanian MSW composition that is shown in fig. 2. The RMSW (biodegradable waste of about 51%) was calculated also taking into account the efficiency of the transfer and sorting as well as composting stations.

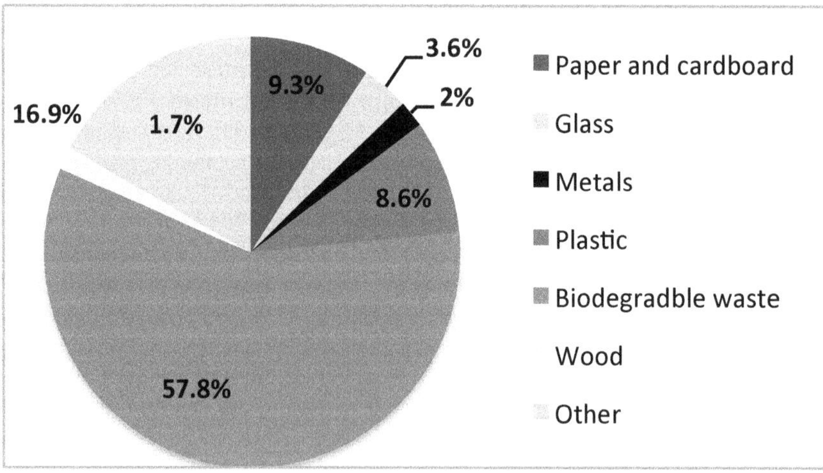

Figure 2: Romanian MSW reference composition.

The lower heating value (LHV) dynamics of the present Romanian RMSW, bio-dried material (BD) and SRF during the bio-drying treatment are reported in fig. 3 [10]. It must be underlined that the LHV increase of bio-dried material and of SRF after two weeks is respectively around 24% and 40%. This is not an energy increase because it must be taken into account that the available mass of fuel is lower after the process. The process allows "concentrating" the initial energy with a contemporary consumption of electrical energy. Generally the energy available at the end of the process is about 3% lower than the initial one [5, 10], apart from the electricity needs that change depending on the adopted technology. The SRF was assessed from the material after the bio-drying process taking out the glass, metals and inert fractions.

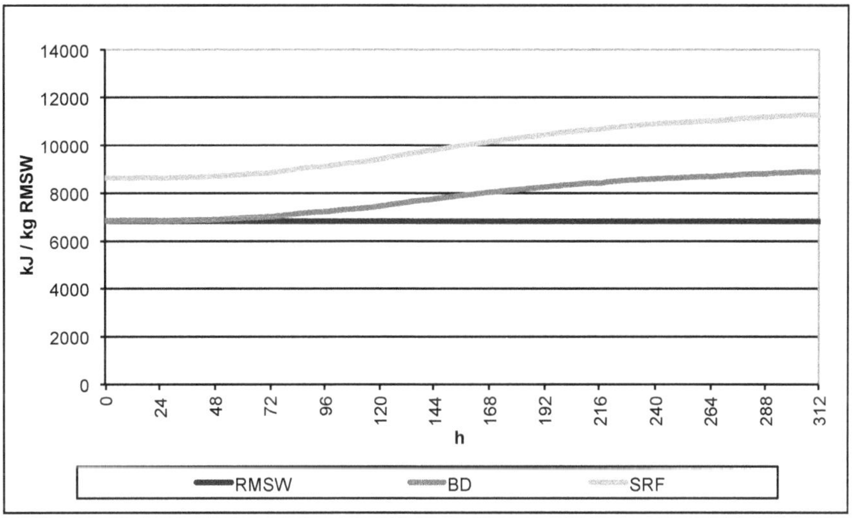

Figure 3: LHV dynamics during the bio-drying process.

The implementation of the principle of waste management and selective collection will change the characteristics of the Romanian MSW and RMSW [24]. Anyway, in spite of the aim to avoid an increase of MSW generation, it must be taken into account that the future *per-capita* production of MSW is expected to increase. In the future scenario, the quantity of materials with high LHV are expected to decrease in the RMSW thanks to the SC of paper, cardboard a packaging waste. If the Take-Back-Programs (TBP) will be also implemented, the residues from the recycling of these fractions with high LHV will be returned to the RMSW [26]. Also the percentage of biodegradable materials and glass are expected to decrease significantly from RMSW thanks to the SC. Taking into account these indications, the future RMSW composition was hypothesized and its LHV was calculated, resulting to about 10 MJ/kg_{RMSW}. The presence of inert in RMSW will depend on the correct management of residues from demolition/construction: they are special waste but often a part of them is disposed of as MSW. If no attention will be put to this aspect, the percentage of inert in RMSW is expected to increase.

Moreover, it must be underlined for the future Romanian RMSW could be assumed as a SRF-like thanks to its LHV, which is similar with the one producible from the current RMSW (about 11 MJ/kg_{SRF}). This aspect was recently discussed in the literature [26] but more studies must be carried out in order to comply also with the respirometric and chlorine and mercury values to be imposed for a correct management [8]. In practice, RMSW bio-drying could be a viable solution for the years of transition towards an optimized MSW management (contributing in the short period to the compliance of the EU Directives regarding the landfilling minimization and the use of renewable sources of energy), but in the long period its adoption could be no longer necessary.

3 Potential of SRF use

The SRF can be used in different thermal plants as indicated in the UNI norms [7, 8]. In table 1, the LHV of different conventional and unconventional fuels that can be used in cement factories are reported. The use of these materials in cement factories as a substitute of coal must be in concordance with the environmental limits for the cement factory process air and also must not increase the percentage of the pollutants in the final products of the cement factory. In EU, mainly in Germany and Austria, the substitution came to overcome the 50% of the traditional fuel. However in these situations very important changes were made to the process air treatment of the plants with an economic investment [10, 27–29].

Table 1: Calorific value of different fuels and fuel-like products.

Fuels and fuel-like products	Calorific value [MJ/kg]
Wood	about 16
Paper and cardboard	3–6
Textiles	> 40
Plastic	17–40
SRF	5–22
Rubber	about 26
Industrial sludge	8–14
Urban sludge	12–16
Fat animal waste	14–32
Coal	20–30
Agricultural waste	12–16
Oil	25–36

Generally, for one kg of concrete, it is necessary to provide about 3,300 kJ/kg of thermal energy in the clinker kiln and between 90 and 150 kWh/$t_{concrete}$ of electric energy [11–13]. For this reason, the use of SRF became in recent years the most attractive solution. Another important aspect connected to the SRF use in the cement factory regards the reduction of greenhouse gas effects as CO_2 and SO_2

thanks to the high temperature in the co-incineration techniques (about 1500°C). Concerning NO and NO_2 (that is, NOx) the high temperature of co-combustion in cement works causes an increase in their emissions. Taking into account that the basic off-gas treatment line of such a plant is simpler than the one of a modern incinerator, the emission factor of NO_x could be worsen. In order to avoid problems with the local air quality, innovative approaches are under discussion, thanks to the availability of NO_2 sensors that can be placed in strategic sites of the territory to act as sentinels [30]. However, these additional costs and the costs for the eventual modification of the off-gas treatment line can be counterbalanced by the fact that a significant amount of coal is saved and the delivery of combustible waste to the plant is regulated by a fee for their treatment.

In Romania, today the co-incineration capacity of the presented cement factory is about 600,000 t_{waste}/y [11–13], treating about 0.08% of the produced waste. The co-incinerated waste were the ones presented in table 1 together with the residues obtained from the treatment of these fractions and not SRF because, at the moment in Romania, a bio-drying plant or other "waste to SRF" plants do not exist but some of them are under discussion.

4 Global and local balances

One of the most important parameters regarding the global environmental balance is the CO_2 emission factor referred to a unit of energy. It is clear that the CO_2 measurable in the bio-drying air is the biogenic one [5, 16] but the operation of a bio-drier needs electricity, with an indirect generation of CO_2. Additionally a correct balance must take into account both the CO_2 from SRF co-combustion and the CO_2 from SRF production. In this respect, only a part of CO_2 from SRF combustion comes from not renewable sources, as RMSW and so SRF are partially biomass.

In table 2, the CO_2 emission from different landfilled and combusted fractions are reported demonstrating that combustion is an interesting pathway from the environmental point of view, referring to greenhouse gases [31].

Similar advantages can be expected for SRF combustion. A case that can be taken into account is the one related to CO_2 emitted using 1 GJ coming from coal or pet-coke or fuel oil or lignite used in industrial plants compared to one 1 GJ from SRF. The CO_2 emission factors for the selected conventional fuels can be

Table 2: CO_2 emission factors comparison [CO_2eq/t].

Fractions	Landfilling	Incineration
Paper and cardboard	2.20	1.3
Plastic	3.1	1.5
Textiles	18	2
Green waste	0.2	-0.12
Food waste	4.5	4.2

taken from the literature [32]. The CO_2 emission factor for SRF can be assessed taking into account the plastics content in the present and future Romanian waste (CO_2 from biodegradable fractions combustion is not taken into account). Moreover the production of SRF is related to electricity consumption depending on the characteristics of MSW (pre-treatment, bio-drying process, post-treatment) [5, 16]. The CO_2 emissions related to the used electrical energy can be referred to a Romanian thermal power plant. Starting from those hypotheses, in fig. 4, a comparison of emission factors is presented [1].

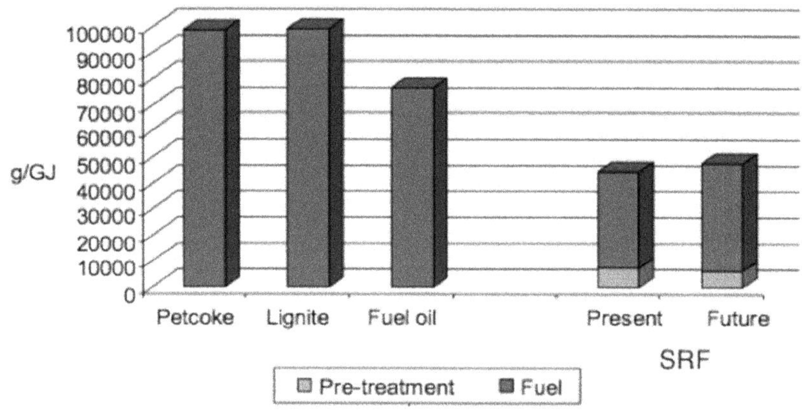

Figure 4: Comparison between CO_2 emission factors.

The fuel substitution has another advantage: if SRF is not produced, it could be supposed that RMSW will be sent to a sanitary landfilling. Romania can postpone RMSW pre-treatment before landfilling until 2017 [20]. The biogas that will be generated gives a significant CO_2 emission factor because of fugitive emissions. Indeed only a part of biogas can be collected, about 50%. The emitted biogas has a CH_4 content of about 60%. In terms of greenhouse gas equivalence it must be taken into account that methane is 21 times more impacting than CO_2 [32]. As a consequence, an emission factor from MSW landfilling can be assessed (always expressed as g_{CO2}/GJ of energy in the fuel/waste) that can reach the same order of magnitude as the ones of conventional fuels. If SRF is produced, this contribution on the global balance is set to zero [5, 16].

Concerning local balances, a question could be related to the expected variation of emissions for instance of heavy metals. An answer could be related to the content of heavy metals in SRF compared to the one of conventional industrial fuel. In this respect, data on heavy metals in German SRF and coal demonstrate that for some heavy metals the environmental balance could be even better using SRF [1].

Concerning the co-generation plants for steam and electricity production in industrial sites for internal uses, a problem to be faced with is the compatibility of SRF with the existing burner: for instance SRF cannot substitute a liquid fuel in a system designed only for that. As a consequence the strategy based on SRF use should be clear from now in Romania as a number of new industrial plants are under design thanks to the attraction of foreign investments.

Another aspect regards the role of the agencies for the environmental control in Romania in order to guarantee that local impact is adequately low. A network already exists. The Romanian agencies already manage a few monitoring stations. The problem is that the conventional approach of air quality monitoring is not oriented to find specific criticalities. Thus, when an authorization to co-combustion is given, a detailed study on the local effects should be promoted. In this respect, particular attention should be given also to the role of waste transport from the environmental point of view: long distance transport of SRF could give unreasonable environmental balances [33].

5 Conclusions

Some considerations can be generated from the present paper:

- the present Romanian RMSW is suitable for an efficient bio-drying aimed to SRF generation;
- SRF can be an interesting substitute of conventional fuels in the industry;
- global environmental balances can be interesting thanks to the partial presence of biomass in SRF;
- SRF can be an alternative to the present use of landfilling;
- an adequate local impact can be guaranteed supporting a correct development of the activities in the local environment agencies.

What must be taken into account is that in a Romanian scenario it could be more favorable to adopt a strategy based on a decentralization of the energy exploitation of RMSW (based on SRF) as it could lower the capital costs compared to the construction of RSW incinerators.

The LHV of the Romanian the fuel obtainable from waste is more or less in line with the ones from the other EU countries where the SC has not yet very high efficiencies. Thanks to this, it is clear that the solutions applied in the other EU countries for the waste valorisation as energy sources can be applied also in Romania. However, Romania has to stimulate the use of renewable energy sources, including waste, as a part of the energy and environmental policies, promoting the use of SRF as alternative fuel through.

References

[1] Rada, E.C., Ragazzi, M. & Apostol, T., Role of refuse derived fuel in the Romanian industrial sector after the entrance in EU. *WIT Transactions on Ecology and the Environment*, **109**, pp. 89-96, 2008.

[2] Rada, E.C., Ragazzi, M., Panaitescu, V. & Apostol, T., Municipal solid waste pre-treatment: A comparison between two dewatering options. *WIT Transactions on Ecology and the Environment*, **102**, pp. 943-949, 2007.

[3] Rada, E.C., Ragazzi, M., Panaitescu, V. & Apostol, T., Experimental characterization of municipal solid waste bio-drying. *WIT Transactions on Ecology and the Environment*, **92**, pp. 295-302, 2006.

[4] Velis, C.A., Longhurst, P.J., Drew, G.H., Smith, R. & Pollard, S.J.T., Biodrying for mechanical–biological treatment of wastes: A review of process science and engineering. *Bioresource Technology*, **100(11)**, pp. 2747-2761, 2009.

[5] Rada, E.C., Ragazzi, M. & Badea, A., MSW bio-drying: design criteria from a 10 years research. *Scientific Bulletin*, series D, **74(3)**, pp. 209-216, 2012.

[6] Velis, C.A., Longhurst, P.J., Drew, G.H., Smith, R. & Pollard, S.J.T., Production and quality assurance of solid recovered fuels using mechanical-biological treatment (MBT) of waste: A comprehensive assessment. *Critical Reviews in Environmental Science and Technology*, **40(12)**, pp. 979-1105, 2010.

[7] Ragazzi, M. & Rada, E.C., RDF/SRF evolution and MSW bio-drying. *WIT Transactions on Ecology and the Environment*, **163**, pp. 199-208, 2012.

[8] Rada, E.C. & Andreottola, G., RDF/SRF: Which perspective for its future in the EU. *Waste Management,* **32(6)**, pp. 1059-1060, 2012.

[9] Ionescu, G., Rada, E.C., Ragazzi, M., Marculescu, C., Badea, A. & Apostol, T., Integrated municipal solid waste scenario model using advanced pretreatment and waste to energy processes. *Energy Conversion and Management*, **76**, pp. 1083-1092, 2013.

[10] Samolada, M.C. & Zabaniotou, A.A, Energetic valorization of SRF in dedicated plants and cement kilns and guidelines for application in Greece and Cyprus. *Resources Conservation and Recycling,* **83**, pp. 34-43, 2014.

[11] http://www.heidelbergcement.ro/studii-de-caz/protectia-mediului/cum-reducem-poluarea-folosind-combustibili-alternativi.html

[12] http://www.holcim.ro/dezvoltare-durabila/avantajele-co-procesarii-deseurilor.html

[13] http://www.lafarge.ro/wps/portal/ro/2_2_2_1-Expertiza_Lafarge_in_utilizarea_deseurilor

[14] Ghinea, C., Petraru, M., Bressers, H.T. & Gavrilescu M., Environmental evaluation of waste management scenarios - Significance of the boundaries. *Journal of Environmental Engineering and Landscape Management,* **20(1)**, pp. 76-85, 2012.

[15] Iacoboae, C., Lua, O. & Petrescu, F., An analysis of Romania's municipal waste within the European context. *Theoretical and Empirical Researches in Urban Management*, **8(4)**, pp. 73-84, 2013.

[16] Rada, E.C., Franzinelli, A., Taiss, M., Ragazzi, M., Panaitescu, V. & Apostol, T., Lower heating value dynamics during municipal solid waste bio-drying. *Environmental Technology*, **28(4)**, pp. 463-469, 2007.

[17] Rada, E.C., Ragazzi, M. & Panaitescu, V., MSW bio-drying: An alternative way for energy recovery optimization and landfilling minimization. *Scientific Bulletin*, series D, **71(4)**, pp. 113-120, 2009.

[18] Negoi, R.M., Ragazzi, M., Aposto, T., Rada, E.C. & Marculescu, C., Bio-drying of Romanian municipal solid waste: An analysis of its viability. *Scientific Bulletin*, series C, **71(4)**, pp. 193-240, 2009.

[19] Rada, E.C., Venturi, M., Ragazzi, M., Apostol, T., Stan, C. & Marculescu, M., Bio-drying role in changeable scenarios of Romanian MSW management. *Waste and Biomass Valorization,* **1(2)**, pp. 271-279, 2010.

[20] Cioranu, S.I. & Badea, A., Different strategies for MSW management in two Romanian cities: Selective collection versus bio-drying. *Scientific Bulletin*, series D, **75(3)**, pp. 151-158, 2013.

[21] www.anpm.ro/Mediu/deseuri-7

[22] Rada, E.C., Ragazzi, M., Panaitescu, V. & Apostol, T Energy from waste: the role of bio-drying. *Scientific Bulletin*, series C, **67(2)**, pp. 69-176, 2005.

[23] Ragazzi, M. & Rada, E.C., Effects of recent strategies of selective collection on the design of municipal solid waste treatment plants in Italy. *WIT Transactions on Ecology and the Environment* **109**, pp. 613-620, 2008.

[24] Rada, E.C., Effects of MSW selective collection on waste-to-energy strategies. *WIT Transactions on Ecology and the Environment*, **176**, pp. 215-223, 2013.

[25] Consonni, S., Giugliano, M., Massarytto, A., Ragazzi, M. & Saccani, C., Material and energy recovery in integrated waste management systems: Project overview and main results. *Waste Management*, **31(9-10)**, pp. 2057-2065, 2011.

[26] Rada, E.C. & Ragazzi, M., Selective collection as a pretreatment for indirect solid recovered fuel generation. *Waste Management*, **34(2)**, pp. 291-297, 2014.

[27] Genon, G. & Brizio, E., Perspectives and limits for cement kilns as a destination for RDF. *Waste Management,* **28(11)**, pp. 2375-2385, 2008.

[28] Kara, M., Gunay, E., Tabak, Y., Durgut, U., Yildiz, S. & Enc, V., Development of refuse derived fuel for cement factories in Turkey. *Combustion Science and Technology*, **183(3)**, pp. 203-219, 2011.

[29] Kara, M., Environmental and economic advantages associated with the use of RDF in cement kilns. *Resources Conservation and Recycling*, **68**, pp. 21-28, 2012.

[30] Rada, E.C., Ragazzi, M., Brini, M., Marmo, L., Zambelli, P., Chelodi, M. & Ciolli, M., Perspectives of low-cost sensors adoption for air quality monitoring. *Scientific Bulletin*, series D, **74(2)**, pp. 241-250, 2012.

[31] https://www.foeeurope.org/sites/default/files/publications/FoEE_SERI_measuring_europes_resource_use_0609.pdf

[32] IPCC/OECD/IEA Revised 1996 IPCC Guidelines for National Greenhouse Gas Inventories. 1997.

[33] Ciuta, S., Schiavon, M., Chistè, A., Ragazzi, M., Rada, E.C., Tubino, M., Badea, A. & Apostol, T., Role of feedstock transport in the balance of primary PM emissions in two case-studies: RMSW incineration vs. sintering plant. *Scientific Bulletin*, series D, **74(1)**, pp. 211-218, 2012.

System analysis for integration of landfill energy production into the regional energy supply

G. Kuplais, D. Blumberga & E. Dace
Riga Technical University, Latvia

Abstract

The waste management solutions with landfills of Latvia vary. Some of them are developed according to the concept of biogas production and use in power plants or cogeneration plants (CHP). The use of biogas in CHP depends on energy consumption and supply and availability of energy consumers. It is more often the case when there are no possibilities to connect heat energy consumers; this creates problems of the low energy efficiency in power plants' operation. A structure of energy user in Latvia is characterized by high energy consumption in households, public and service sector, compared with relatively low consumption in the rural and industrial sector. This situation necessitates special tasks for power sector development, especially choosing energy resources to ensure energy production and supply. In Latvia imported fossil and renewable energy resources are used. A use of specific energy resource depends on an energy supply policy, and total consumption of energy resources depends on the development of every type of energy resources in regions. Now there is an unjustified high proportion of fossil fuel in state energy balance which can be reduced by beginning the active use of local fuel in regions. RDF and landfill gas as a renewable energy source could play a role in the development of local energy resources in Latvia, which should be based on principles of cluster, integration and gradualness. In this paper the main attention is paid to waste management problems from the development of a regional power supply system point of view together with understanding, that those are not two separate systems, but two parts of one system with dynamic rather than static development.
Keywords: biogas, landfill gas, refuse derived fuel, regional energy supply, renewable energy sources, system analysis.

1 Introduction

Centralization of waste dumps was considered to be the main task of waste management system of Latvia at the end of 1990s and the beginning of this century, therefore a policy, whereby 500 small scale dumps were supposed to be replaced by 11 big scale landfill sites, was enforced. At that time an idea of large landfill site depositories, where biogas is produced by decomposing organic waste, was implemented in the policy. There were biogas collecting systems developed and established. In regions this conception gave the opportunity to produce energy from biogas.

Problems began when biogas utilization in cogeneration plants (CHP) was encumbered by the lack of possibility to find heat energy consumers, which in turn resulted in low efficiency landfill power plants almost all over Latvia. The emerging situation forced the search for another solution related to biogas utilization in providing the other needs of consumers.

Due to high electricity feed in there is an economical motivation for power plant operation with low efficiency. For electricity produced in renewable energy power plants with a nominal capacity of up to 4 MW high feed in tariff has been transposed in Latvia's legislative acts.

Now, because of the above mentioned reasons, development of Latvia's landfill sites is at the crossroads. On the one hand, it is economically feasible to operate CHP just for electricity production, but on the other – it is important to use natural resources at full value by producing from biogas the maximum amount of heat energy. In the first case, it means that there is no need for waste sorting in landfills, but in the other it is important to sort both – before waste collection and in landfills.

In this paper, the main attention is paid to waste management problems from the development of regional power supply system point of view together with understanding that those are not two separate systems, but two parts of one system with dynamic rather than static development.

In this paper, an analysis of the system is performed based on the operational study of one of Latvia's landfills – landfill "Daibe".

2 Energy production from waste

If it is taken into account that waste is any substance which the holder discards or intends or is required to discard [1], municipal solid waste is a waste type that includes predominantly household waste with sometimes the addition of commercial wastes collected by a municipality within a given area. Municipal solid waste landfill sites are a large source of human-related methane emissions, therefore, if it is not dealt with in a sustainable manner, it can contribute to greenhouse gas emissions and, by implication, climate change.

Latvia has been divided into 11 waste management regions for industrial and municipal waste of which North Vidzeme Region is the second largest by population (see fig. 1). In 2004 municipal waste landfill "Daibe" was put into operation, and since then, is the place for treatment and depositing solid municipal waste in North Vidzeme Region.

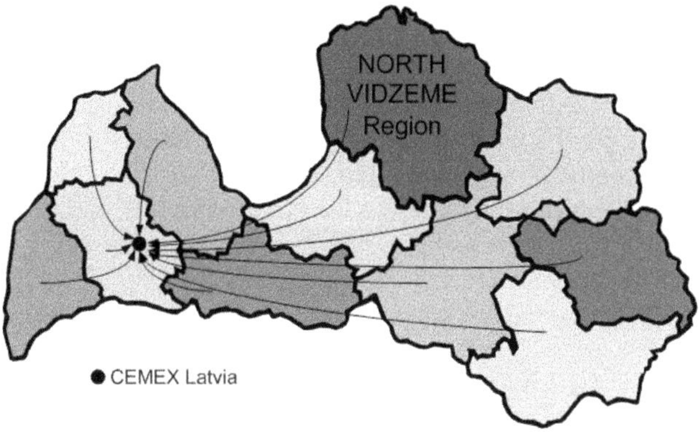

Figure 1: Waste management regions in Latvia with potential distribution of the refuse derived fuel (RDF).

It is obvious that many waste streams are not subject to direct processing and will require special measures in the form of specific pretreatment of the waste prior to processing (see fig. 2). One form of pretreatment is separation and recycling of waste components thereby removing a portion of the waste stream for recycling and other uses [2].

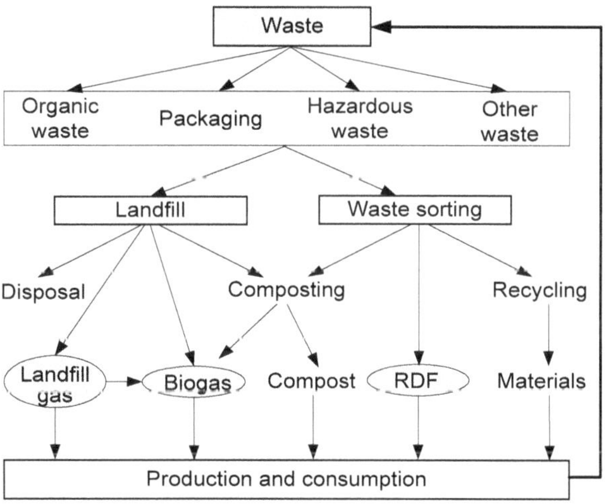

Figure 2: Waste stream processing options.

Municipal and industrial waste could provide feedstock for a number of conversion systems, all of which could recover useful energy while reducing the requirement for landfill sites.

In Latvia the potential of energy efficient utilization of landfills' energy sources is related to waste sorting and RDF production, as well as to organic waste composting with biogas production, utilization and recycling in landfill.

3 Landfill gas

Instead of allowing landfill gas to escape into the air, it can be captured, converted and used as an energy source for existing landfills. Using landfill gas helps to reduce odors and other hazards associated with landfill gas emissions, and it helps to prevent methane from migrating into the atmosphere and contributing to local smog and global climate change [3].

Power generation from landfill gas offers major air quality benefits where landfills already exist or where the decision to build the landfill has already been made [2].

Composition of the landfill gas varies with age and type of waste. Landfill gas is composed of a mixture of hundreds of different gases. By volume, landfill gas typically contains 45% to 60% methane and 40% to 60% carbon dioxide. Landfill gas also includes small amounts of other compounds [4].

Fig. 3 shows the amount of landfill gas emissions in landfill "Daibe", starting from year 2005 up to 2013 (projection). About 50% of it can be collected with the existing technology in the landfill. The measurements of the collected landfill gas show average concentration of methane around 60% in it. The produced landfill gas is incinerated in a cogeneration plant (CHP) with an efficiency factor of 0.37. Fig. 4 shows the amount of electricity produced in this CHP during a 7 month period.

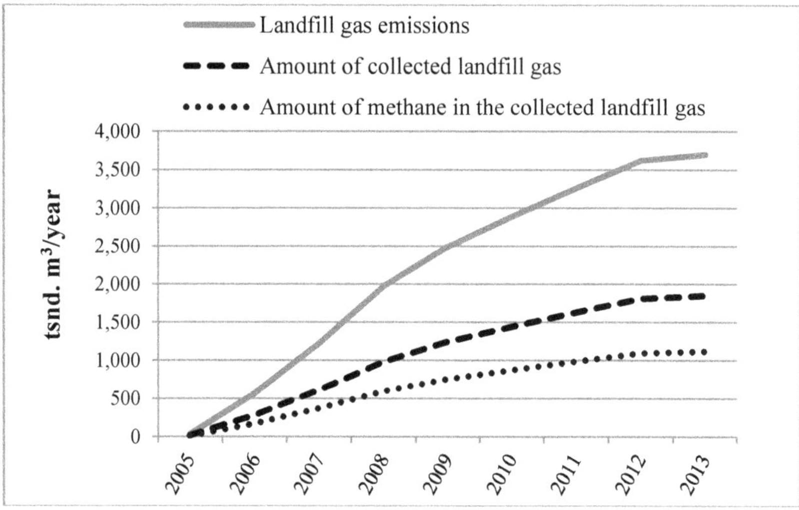

Figure 3: Landfill gas emission and collection parameters in landfill "Daibe".

As can be seen from the graph in fig. 4, the amount of electricity produced changes very rapidly. Mostly it is dependant on the amount of collected landfill gas, however it may be associated also with the methane content in the gas. Another associated problem, which was logged during the CHP operation period (214 days), is that the CHP stopped for 298 times (!).

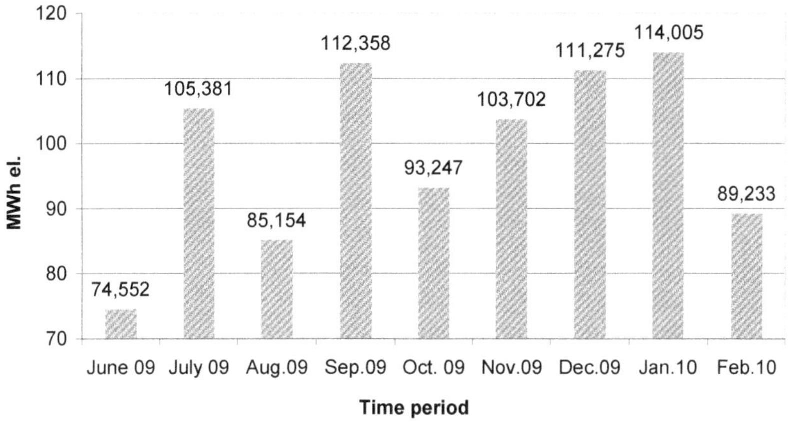

Figure 4: Electricity produced in CHP of landfill "Daibe".

Since methane content is so low, the production of alternative fuels from landfill gas in landfill "Daibe" is now an emerging area. One of the options is biogas production.

4 Biogas production and utilization for reconstructed landfills

Biogas from landfills as a renewable energy source could play a role in the development of local energy resource in Latvia, which should be based on principles of cluster, integration and gradualness. The first principle is based on a situation when regions with landfills select the development of its own energy sector, integrating development plan of energy supply system into territorial plan of area or city. By use of the integration principle (including biogas from landfills), it is possible to forecast total development of state power and energy sector, taking down regional limits of energy supply. A renewable energy grid should be established for use of biogas from landfills, streams of wood, straw and other biomass in the energy sector of Latvia.

Some of the waste management solutions for landfills of Latvia are developed according to the concept of low grade biogas (landfill gas) production and use in power plants or CHPs. The use of biogas in CHP depends on energy consumption and supply as well as availability of energy consumers. More often there are cases when it is not possible to connect heat energy consumers, which create problems of low energy efficiency of power plants' operation.

Other alternative biogas utilization options are injection into natural gas grid or transport fueling.

For the use of biogas as transport fuel or its injection into a natural gas grid, the only suitable fraction in biogas is methane. Its other components (e.g. carbon dioxide, hydrogen sulphide, hydrogen) are unsuitable for these purposes, undesirable or even hazardous; therefore, they should be removed from biogas by cleaning and upgrading it (see fig. 5).

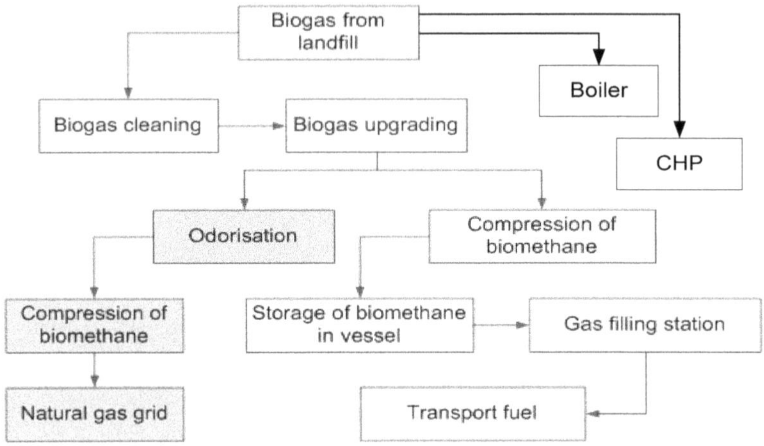

Figure 5: Stages of upgrading the biomass quality.

The cleaning process involves separation of the cumbersome compounds from biogas as well as water vapor separation or biogas moisture content reduction. Upgrading of biogas is a cleaning process when the ballast is drained from it; thereby the methane content of biogas is increased. Methane concentration of landfill gas which is usually 45–60% is increased to over 95%. To reach necessary Wobbe index of gas, carbon dioxide has to be removed.

Certain requirements have to be fulfilled for biogas injection into natural gas grid: specific odors used in the natural gas system have to be added to biogas (process of odorisation) and the pressure has to be increased to match that in the respective section of natural gas grid.

For biogas use in vehicles, a storage and filling system, which ensures effective biogas supply to vehicles, has to be developed.

The landfill "Daibe" is considering the option of upgrading the biogas for its utilization as a transportation fuel. The option of injecting the biogas into natural gas grid is not envisaged for now because of existing barriers, such as high quality requirements for biogas, large investments and lack of state's regulatory normative acts.

5 RDF

One form of materials recycling is the refuse derived fuel (RDF) – form of the combustible fraction of municipal solid waste. In order to produce such a fuel, an RDF plant seeks to concentrate the combustible fraction of the waste by removal of non-combustible materials, such as glass and metals. The conversion of municipal solid waste to RDF depends on the type of process used to produce the RDF; the conversion rate can vary between 23 and 50% by weight. The raw waste has a typical calorific value of 9.1 MJ/kg whilst the processed coarse RDF has a calorific value of approximately 18.0 MJ/kg [5].

The combustion of RDF in cement kilns is a significant route of energy recovery utilization. In case of sorting of waste by production of RDF in Latvia, utilization as energy source will be possible in a cement kiln. CEMEX Latvia – the only cement kiln in Latvia and the largest one in the Baltic States, is ready to use RDF for supplying their production processes with energy. It is estimated that to cover all their production needs with RDF, three times as much waste would be needed as there is in Latvia. It means that even by using all of Latvia's waste for RDF production, it would not be enough to supply CEMEX Latvia.

6 System dynamics analyses

All the above mentioned indicates that the development and major role of landfills in energy supply system of Latvia's regions is dependent on various factors, like, among others, waste management parameters (such as waste sorting grade in landfills and before their transportation to landfills, level of waste recycling), economical factors (such as waste transportation costs, natural resource tax, energy resource costs, energy production and feed in tariffs, development and costs of biogas upgrading methods), impact on decrease of global climate change (such as decrease of landfill gas losses from depositories, decrease of greenhouse gases, decomposition of organic substances and biogas production in special landfill bioreactors).

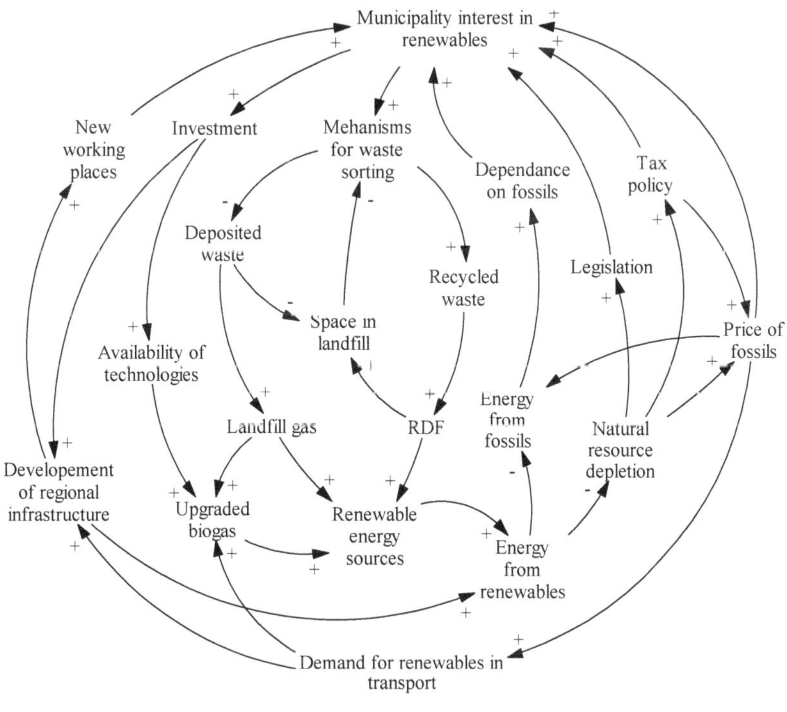

Figure 6: Causal loops' diagram of system dynamics analyses.

System dynamics is a modelling methodology, which permits an analyst to represent graphically and mathematically the interactions governing the dynamic behavior of complex socio-economic systems [6]. The purpose of a system dynamics intervention is to identify how structure and decision policies generate a system behavior identified as problematic, so that structural and policy oriented solutions can be identified and implemented [7].

The analysis of impact of the above mentioned factors is possible by using a system dynamics model, the example of which is illustrated in fig. 6.

The diagram shows that energy produced from renewable energy sources will rise only when interest of municipalities in such an energy will increase. That can happen when renewables promoting legislative acts or taxation policy is in force, as well as growth of dependance on fossil fuels and their price. When that happens, municipalities develop mechanisms and technologies for renewables by investing financial resources.

The decrease of acumulated waste amount in landfill "Daibe" by energy production and integration in the regional energy supply has been analyzed, based on various aspects and using the existing and projected data.

7 Results

The system is analyzed over a 10 year time period – from year 2003 up to year 2013 (projection). Such a time range has been chosen because of accessibility to both past and projected data.

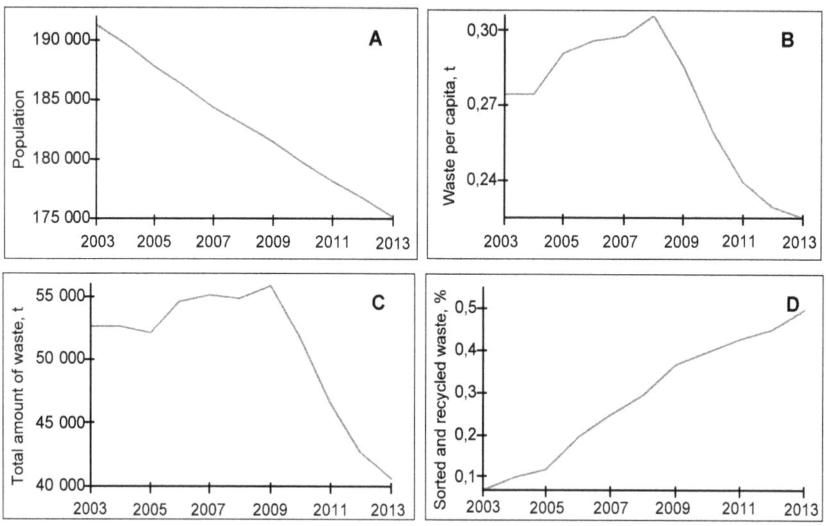

Figure 7: Graphs characterizing North Vidzeme Region with respect to waste production: A – Resident population in the region; B – Amount of waste produced per capita; C – Total amount of waste produced in the region; D – percentage of sorted and recycled waste amount.

Fig. 7 shows the resident population and the amount of waste produced per capita in North Vidzeme Region, as well as the total waste amount produced and the percentage of sorted and recycled waste amount out of it in the region.

It can be seen from the graphs that population is decreasing linearly, whereas the amount of waste produced per capita rises at the beginning, but afterwards decreases quite rapidly. This can be explained by the rapid GDP decrease in the state due to the global economical crisis, which has hit Latvia quite strongly.

The parameters shown in fig. 7 are taken into account to estimate the amount of waste that could reach the landfill. Landfill gas or RDF can be produced from this waste for utilization in energy production. If it is assumed that all the combustible fraction of waste is removed to produce RDF, which is used for energy production in cement kiln of CEMEX Latvia, than only the non-combustible and non-recyclable fraction is deposited in the landfill. In landfill "Daibe" up to 40% of the introduced waste could be considered as non-combustible and non-recyclable fraction. The rest may be used for RDF production.

The dynamics of waste accumulation in the landfill "Daibe" can be drawn by modeling the implementation of waste sorting for the production of RDF. Fig. 8 shows a comparison of the waste accumulation dynamics, if RDF production is implemented this year already and if it is not implemented at all.

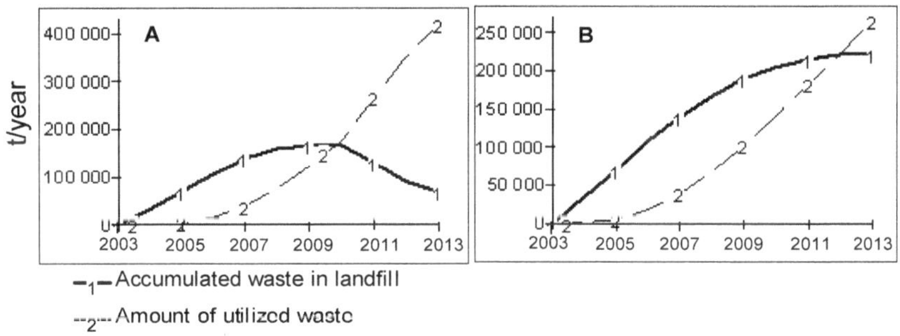

Figure 8: The dynamics of waste accumulation and utilization in the landfill "Daibe": A – RDF production is implemented this year; B – RDF production is not implemented at all.

In fig. 8, graph A shows that the amount of accumulated waste drops rapidly, when waste utilization for RDF production is introduced. If the current system is not changed, that is, the waste is accumulated and only the landfill gas is collected, then the dynamics of waste accumulation looks like the one in graph B – it shows goal-seeking tendency of waste accumulation, and besides that the amount of accumulated waste in 2013 is about 65% more than in the case of RDF production.

System analysis shows that there is great potential for waste utilization in energy production if municipalities are interested in it and if there is enough motivation for it, like legislative acts promoting renewable resource utilization or interest from consumer side. The most promising way to decrease deposited waste in landfill is predicted to be RDF production.

8 Conclusions

1. Waste landfill sites have high RDF potential, which lets its concentrating in cement kiln. In Latvia it is a future solution.
2. System dynamics analysis allows predicting the impact of different factors on waste management and regional energy system of Latvia showing that they have to be considered as one.
3. The example of system dynamics analysis of landfill “Daibe” shows that the amount of deposited waste can be decreased by implementing RDF production. If it is done this year already, than by year 2013 the amount of deposited waste is decreased by 65% compared with the future projections of the existing waste management system.

References

[1] Directive 2006/12/EC of the European Parliament and of the Council of 5 April 2006 on waste // Official Journal of the EU, L 114/9 (27.4.2006.)

[2] Speight, J., *Synthetic Fuels Handbook*, McGraw-Hill: New York, pp. 323–365, 2008.

[3] Barlaz, M.A. & Ham, R.K., Leachate and gas generation. *Geotechnical Practice for Waste Disposal*, ed. D.E. Daniel, Chapman and Hall: London, Chapter 6, pp. 113–136, 1993.

[4] Landfill gas basics, *Landfill Gas Primer*, An Overview for Environmental Health Professionals, Department of Health and Human Services, Agency for Toxic Substances and Disease Registry, Division of Health Assessment and Consultation, pp. 3–12, November 2001.

[5] Williams, P.T., *Waste treatment and disposal*, John Wiley and Sons: Hoboken, pp. 127–171, 2005.

[6] Vizayakumar, K., Environmental policy analysis: System dynamics approach. *Public policy analysis and design*, ed. V.K. Agnihotri, Concept Publishing Company: New Dehli, Chapter 20, pp. 311–329, 1995.

[7] Saleh, M., Oliva, R., Kampmann, C.E. & Davidsen, P.I., A comprehensive analytical approach for policy analysis of system dynamics models. *European Journal of Operational Research*, **203**, pp. 673–683, 2010.

Selective collection and waste-to-energy strategies: which interaction?

E. C. Rada
University of Trento, Italy

Abstract

This paper updates and analyses the interaction between the selective collection of municipal solid waste and waste-to-energy strategies in a case-study where high efficiencies of source separation have been reached thanks to more than a decade of efforts. For a better understanding of the local waste management history, in the selected case-study (province of Trento, Italy) the most important years were analyzed and characterized. In the selected case-study, particular attention has been given to the organization of selective collection and environmental protection, the location being also a tourist area. The variability of selective collection (that increased its efficiency year by year) affects the characteristics of the residual municipal solid waste and subsequently its "lower heating value" and the waste-to-energy strategies. The residual municipal solid waste generated today is similar to a solid recovered fuel. Its exploitation in existing industrial plants for energy recovery represents one of the present-future strategies to be developed.
Keywords: MSW, RMSW, selective collection, SRF, waste-to-energy.

1 Introduction

In recent years, municipal solid waste (MSW) has become one of the most important indicators of the socio-economic development of a country [1]. The present paper updates and analyses the interaction between the selective collection of MSW and waste-to-energy in an area in the north part of Italy [2].

Italy is one of the co-founding countries of the European Community, which reached a 35% efficiency of selective collection (SC) at national level in 2013 [3]. It is clear that duration of membership is not correlated with the SC organization that depends on different visions and priorities of waste management (France and Belgium, co-founding countries, have SC under 40%; Slovenia, instead after only

7 years arrived to 45%). However, among the 28 countries, only 3 northern countries exceeded 50% of the SC target at national level (Germany, Luxembourg, Netherlands, Sweden) and in the others, the SC has very variable efficiency (1–45%) [4].

The first concept of waste selective collection (kerbside) and open dumps constructed at about 2 km away from the villages, date back to the birth of the first civilization in 2000 B.C. [5]. After the war and with the economic-industrial development and consequent pollution, some environmental limits and indications regarding waste collection, treatment and landfilling (422/75/EC, 94/62/EC) were developed and implemented.

The first proposed treatment for the selected organic waste was the composting one [6], followed by anaerobic digestion [7]. After few experiences around the world, in the 60s the incineration was massively proposed in order to treat the various kinds waste [8]. Since 2000, the mechanical biological treatments (MBT) were widely implemented for MSW [9, 10]. In recent years the MSW sector paid more and more attention to the energy optimization and environmental impact minimization of the above-mentioned processes [11–14].

Recently, SC increased its role in many countries for many fractions [15–19] as an integrated option of the conventional processes, showing an important impact on the management of the residual MSW (RMSW). The variability of SC affects the characteristics of the RMSW and subsequently its lower heating value (LHV) and the waste-to-energy strategies. The RMSW in the areas where SC has high efficiency could be considered as solid recovered fuel (SRF) [20].

In this context, the aim of this paper is to analyze the consequences of the evolution of the MSW selective collection towards the strategies of waste-to-energy, referring to an Italian case-study, where particular attention has been given to the organization of SC and environmental protection starting from 2000.

2 Materials and methods

The Trentino region was selected as case-study in order to give an overview of the waste SC development over the years and the consequences of its implementation on waste-to-energy strategies. In 2013, in the selected case-study, the inhabitants were about 610,000 (15% were assumed as inhabitants equivalent thanks to the tourism influence) and produced about 265 kton of MSW [21]. The territory involved is characterized by a significant richness of data in the MSW sector. This is the consequence of a political choice that allowed opening a specialized agency for "observing" the waste sector in order to maximize mass valorization before facing the implementation of a waste-to-energy plant.

The first step in order to improve the MSW management in the chosen province was made through the regulation L.P. 29/82 that forbade the open dumps (300 in province in 1982) and imposed the sanitary landfills (7 controlled landfills were realized in the following years). However, the MSW collection and transportation to a final disposal was given to the municipalities since 1941, thanks to the legislation 366/1941.

In 1993, the first disposal plan for MSW was developed (SC efficiency at the beginning was near 4%), followed by updates: 1997, 2002 and 2006. In the following years, a continuous update of data on waste was performed (SC efficiency reached 75% in 2013).

In the year 1998, the same province issued the L.P. n. 5/1998 regulation in order to give information about the organization and the development of SC and about how to minimize the MSW landfilling.

The consequences can be seen in the data reported in the waste management plans issued in the last two decades: data from the years 1994, 2001, 2004, 2010 and 2012 were analyzed in order to extract the information needed for characterizing the local dynamics of the MSW sector. At the beginning almost all the collected waste was going to landfills with and without biogas recovery and only a very small part of the biodegradable waste was used for compost production. The situation changed significantly thanks to the German Topfer Decree in 1991 that enforced the concept of the material recovery facilities (MRF). As a consequence, in the chosen province, the MSW collection was made through three bins: biodegradable, recyclable and residual waste.

In fig. 1, the composition of MSW is shown pointing out the present situation (first bars) and comparing it to that of the previous years.

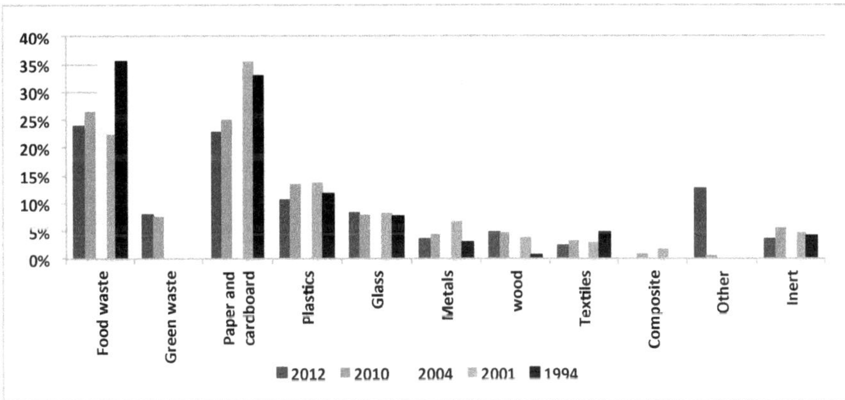

Figure 1: MSW composition during the last two decades.

In 2001 and 1994 green waste was not classified; this is the reason why the 1994 bar for food waste is the highest. The scenario in 2001 was similar in terms of MSW composition but the enforcement of the Italian Ronchi's Decree (22/1997) accelerated the evolution of SC, also including the food waste collection. The SC was performed through home collection (kerbside), street collection and eco-stations, for specific fractions or mixed fractions [17, 20]. This system is used also today. In table 1, an overall view of the SC system in 2001 is reported for the 11 municipalities of the chosen province. In some municipalities, multimaterials (glass, plastic, can) are collected together, in others one by one. The RMSW were sent to 8 landfills and in two of them, the RMSW was added as bale.

Table 1: SC collections strategies.

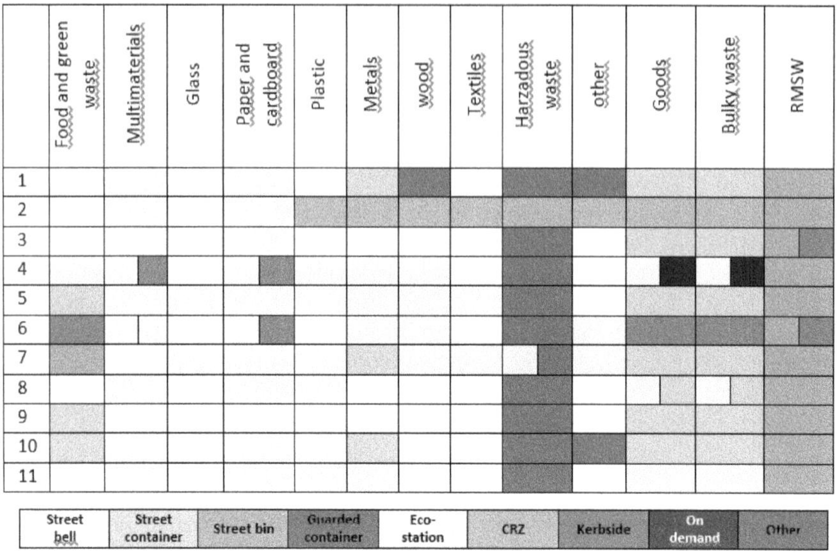

In 2004 the construction of a centralized incinerator for the valorization of the residual MSW (RMSW) was under discussion (its project was analyzed and modified over the next decade). The SC of materials with high biodegradability and moisture content increased reaching about half of the available steams.

In 2010, the incinerator construction was still delayed. In the meanwhile, the management of the growing streams of food and green waste selectively collected was associated to the planning of 4 anaerobic digesters (one of them started its activity in 2012; a second one was completed, but its full operation was delayed because of the need of an optimization of the recirculation streams) mostly coupled with post-composting after biogas exploitation. The obtained results in diverting biodegradable materials from landfilling are in agreement with the targets of the EU landfilling directive.

A new regulation for the solid recovered fuel (SRF) from MSW was implemented in Italy in 2010 [20, 22, 23]. It reopened, in 2012, the discussion on the use of RMSW as a SRF-like in co-combustion plants already existing also out of the province.

In 2013, it was decided to exclude the incinerator from the waste management plan. A second digester available was operated partially, but a full performance is expected in 2014. From the economic point of view, the amount of RMSW to be treated is going to be too low for such a plant. The new vision includes bio-drying and post-treatment for SRF production as local option for RMSW management [20, 22, 23]. The SRF solution was proposed also in order to recover combustible materials from a landfill where about 10^5 waste bales were stored.

In spite of a near steady composition of MSW, the variation of the RMSW characteristics as a result of SC dynamics has changed the strategies for its treatment. Useful parameters for analyzing this evolution are the lower heating value (LHV) of RMSW, the rate of exploitable energy from MSW (as RMSW) and the percentage of food waste in RMSW. All these parameters were assessed for the five years of reference.

3 Results and discussion

Fig. 2 shows data related to each of the considered years, concerning the dynamics of the RMSW composition. The decrease of food waste percentage as consequence of source separation is clear. A higher LHV of RMSW has to be expected. The implementation of the limits requested by the regulation 22/97 (15% SC efficiency in 1999, 25% in 2001, 35% in 2003, and for the following years, 40% in 2005 and 50% in 2007 respectively) changed significantly the composition of RMSW as can be seen from fig. 2.

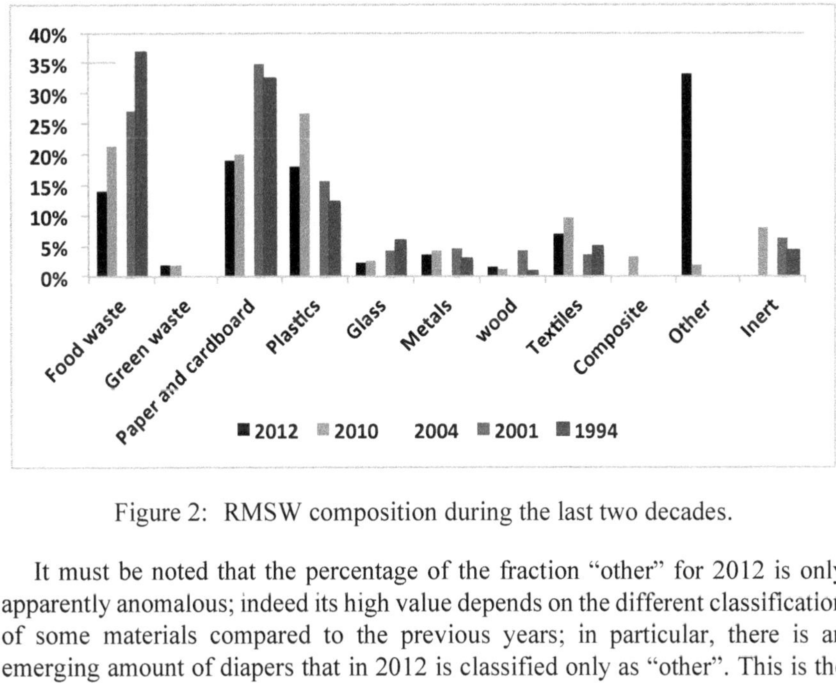

Figure 2: RMSW composition during the last two decades.

It must be noted that the percentage of the fraction "other" for 2012 is only apparently anomalous; indeed its high value depends on the different classification of some materials compared to the previous years; in particular, there is an emerging amount of diapers that in 2012 is classified only as "other". This is the reason why the fraction "composite" is zero in 2012.

Fig. 3 shows the trend of SC of the main streams. The evolution of the citizens' behavior is remarkable: for instance, about 20 years ago the rate of source separation for paper and cardboard was around 5%; today it is higher than 75%. Food waste segregation changed too: from zero to around 75%. A target of 85% is expected for this last stream. Applying this value to a generation of 60,000 t/y and assuming a conversion rate of 0.080 Nm^3 of CH_4 per kg of food waste fed in the planned digesters, the potential production of methane in the province is around 4 millions of cubic meters per year. An efficiency of electrical conversion as 40% could give a yearly generation of electricity around 16 million of kWh_{el}.

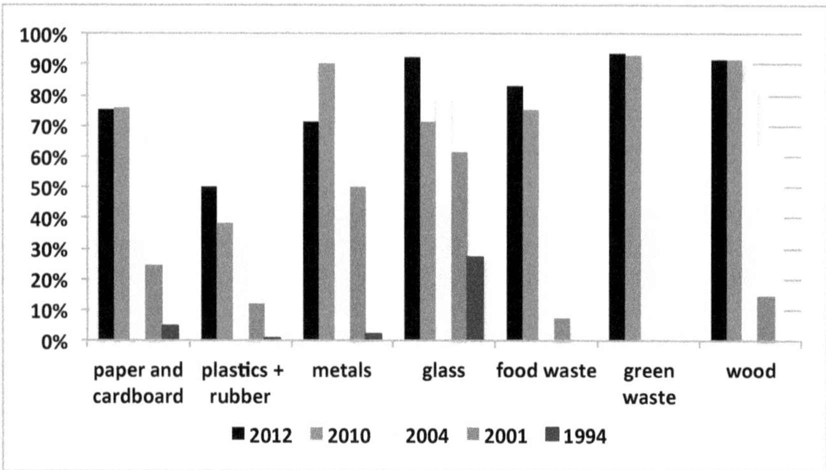

Figure 3: SC during the last two decades (percentage of interception).

The citizens' effort allowed reaching overall efficiencies as follows: about 4% in 1994, 16% in 2001, 32% in 2004, 61% in 2010, 67% in 2012 and 71% in 2013 respectively. After a slight delay in complying with the Ronchi's targets, a change of attitude can be observed after 2004: the efficiency doubled in only 6 years. One explanation can be related to the organizing efforts of the local authorities that set specific initiatives for decreasing the amount of RMSW to be sent to the proposed incineration. After 2010 additional percentage points were obtained thanks to refining actions.

In fig. 4, data on LHVs of MSW and RMW are reported for the five years. The present situation points out how SC has recently increased the LHV of RMSW compared to the one of MSW. Some comments can be made:

- The initial value of LHV (1994) was already adequate for direct combustion; this had to be expected as the studied area was classified as high income economy still in the 90s;
- The "concentration" of energy in the RMSW of 2010 cannot be the unique parameter to be taken into account for analyzing the energy viability of the present scenario; indeed another aspect to be considered is the ratio between the energy available in 1 ton of MSW and the energy available in the respective amount of RMSW;
- It is a general opinion that SC causes an increase of LHV of RMSW; in reality, the available data demonstrate that this trend is approximately visible from 1994 to 2010 (apart from the fluctuation in 2004), but after that year, the extreme source separation of light packaging made the energy content of RMSW poorer;
- The LHV of MSW in 2012 is unexpectedly lower than the one in 2010; this can be partially explained by the higher percentage of "other" fraction, with consequent difficulty in assessing its LHV.

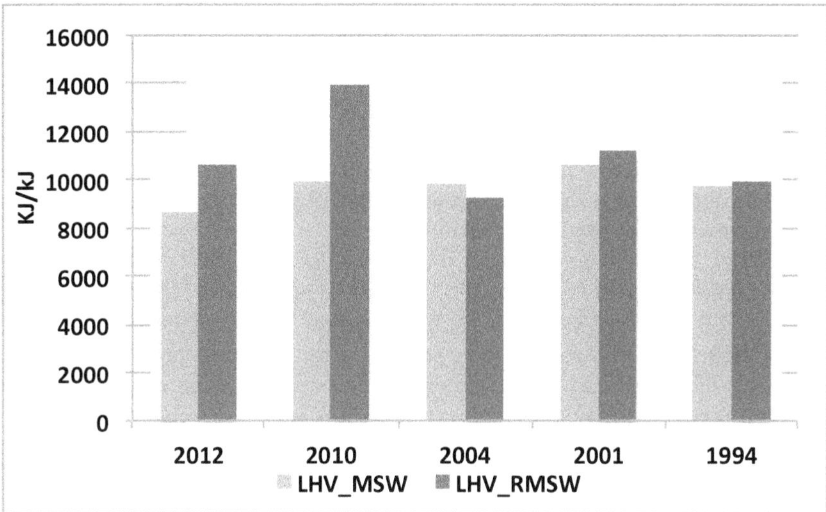

Figure 4: LHV of MSW and RMSW during the last two decades.

The energy ratio is presented in fig. 5 where it is clear that the higher SC is, the lower the energy exploitability rate, as a consequence of the segregation of waste fractions with a positive energy content (plastics, paper and cardboard in particular). Taking into account the MSW per capita amount of 2010 (469 kg inh^{-1} y^{-1}), the SC rate (65%), the population with tourists expressed as resident equivalents (574,373) and the LHV of RMSW (13,939 kJ/kg), an

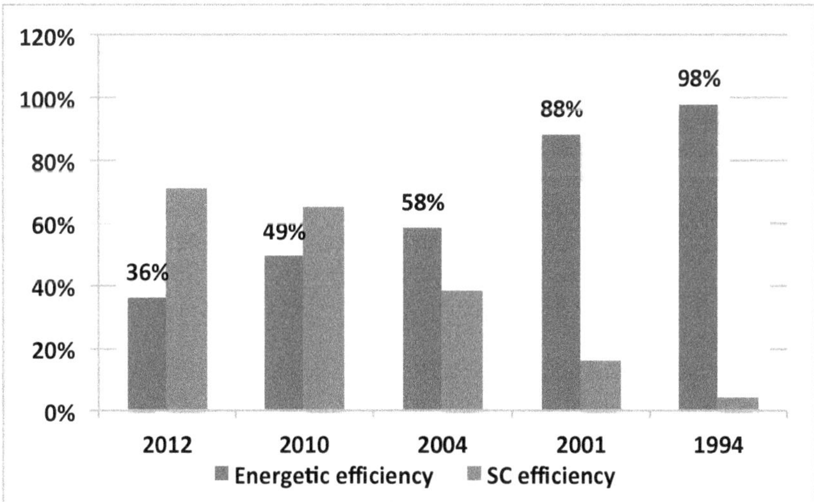

Figure 5: Energy exploitability (RMSW vs MSW) and SC efficiency.

expected net electricity generation from incineration as 25%, the electricity generable from RMSW should be around 91,300,000 kWh_{el}/year (about 5 times higher than the one expected from the digesters). Data related to the year 2012 do not confirm this forecast as both the amount of RMSW and its LHV are lower. This instability was one of the topics in the local discussion about the availability of this option. If there is not a guaranteed input for a waste-to-energy plant, the tariff to be applied "at the gate" becomes unstable. This is one of the reasons why the local strategy moved towards SRF.

The 2012 RMSW characteristics (high LHV, low food waste content, overall amount significantly lower (about 70,000 t/y), expected low Hg and Cl content [17, 20, 22]) recently opened a local discussion on the viability of a direct classification of RMSW as SRF to be sent "directly" into existing plants also out of the provincial territory. This is a demonstration of how an extreme SC can affect the Waste-to-Energy strategies in the MSW sector.

4 Conclusions

SC can cause significant indirect modifications of RMSW that can become similar to a SRF, opening to near direct co-combustion opportunities in existing industrial plants (cement factories, thermal power plants, etc.). This pathway could decrease the cost of solutions alternative to direct combustion in dedicated plants (incinerators).

In spite of its high LHV, the energy exploitable, compared to MSW, is significantly reduced as a consequence of SC. That decreases the incineration scale to a value that must be well analyzed in terms of economic viability. Indeed incineration is affected by a significant scale effect that is more relevant to low plant capacity. The parallel increase of food waste SC opens to interesting energy recovery by anaerobic digestion. Of course the viability of this strategy needs a steady cooperation of the citizens that must guarantee an efficient separation of food waste. Data presented for the selected case-study demonstrate that when an organization effort is well performed, the acceleration of SC is remarkable, with significant consequences on the overall strategy of MSW management.

From the above-presented scenarios, it is clear that the MSW management evolved to a waste-to-energy configuration (thanks to the direct and indirect generation of biogas and SRF-like) integrated with mass valorization (thanks to the SC optimization).

An open question concerns the quality of the source separated waste streams. The improvement of their quality is one of the main targets of the future MSW management in the analyzed province. The first tool adopted to this concern can be considered the introduction of the punctual tariff (each user pays according to the RMSW that generates). This topic will be the subject of a future article.

Acknowledgements

The author wishes to acknowledge PAT (Provincia Autonoma di Trento – *Autonomous Province of Trento*, Italy) and Mr. Zatelli, in particular, for the support in data availability.

References

[1] McCollough, J., Determinants of a throwaway society – A sustainable consumption issue. *Journal of Socio-Economics,* **41(1)**, pp. 110-117, 2012.
[2] Rada, E.C., Effect of MSW selective collection on waste-to-energy strategies. *WIT Transaction on Ecology and the Environment*, **176**, pp. 215-223, 2013.
[3] Consonni, S., Giuliano, M., MAssarutto, A., Ragazzi, M. & Saccani, C., Material and energy recovery in integrated waste management systems: Project overview and main results. *Waste Management,* **31(9-10)**, 2057-2065, 2011.
[4] http://appsso.eurostat.ec.europa.eu
[5] CAC – Cooperativa Atlantide di Cervia, Progetto scuola città, Rifiuti: Energia del futuro, 1998, www.atlantide.net
[6] Gotaas, H.B., Composting; sanitary disposal and reclamation of organic wastes. Monograph series. *World Health Organization,* **31**, pp. 1-205, 1956.
[7] Johnson, G.E., Kunka, L.M., Decker, W.A. & Forney, A.J., Production of methane by the anaerobic decomposition of garbage and waste materials. *American Chemical Society, Division of Fuel Chemistry*, **16(4)**, pp. 70-78, 1972.
[8] Dennis, R., Muller, F.L., Kristal, E. & Silverman, L., Special incineration studies – institutional design. *ORINS Reports. U.S. Atomic Energy Commission 7593*, pp. 344-364, 1960.
[9] Adani, F., Scatigna, L. & Genevini, P., Biostabilization of mechanically separated municipal solid waste fraction. *Waste Management and Research,* **18(5)**, pp. 471-477, 2000.
[10] Rada, E.C., Ragazzi, M., Panaitescu, V. & Apostol, T., Experimental characterization of municipal solid waste bio-drying. *WIT Transactions on Ecology and the Environment*, **92**, pp. 295-302, 2006.
[11] Cadena, E., Colon, J., Artola, A., Sanchez, A. & Font, X., Environmental impact of two aerobic composting technologies using life cycle assessment. *International Journal of Life Cycle Assessment,* **14(5)**, 401-410, 2009.
[12] Rada, E.C. & Ragazzi, M., Critical analysis of PCDD/F emissions from anaerobic digestion. *Water Science and Technology*, **58(9)**, pp. 1721-1725, 2008.
[13] Rada, E.C., Ragazzi, M., Zardi, D., Laiti, L. & Ferrari, A., PCDD/F environmental impact from municipal solid waste bio-drying plant. *Chemosphere*, **84(3)**, pp. 289-295, 2011.
[14] Ragazzi, M. & Rada, E.C., Multi-step approach for local environmental impact comparison among conventional and innovative MSW thermo-chemical treatments. *Chemosphere,* **89(6)**, pp. 694–701, 2012.

[15] Torretta, V., Ragazzi, M., Istrate, I. & Rada, E.C., Management of waste electronic and electronic equipment's in two EU countries: a comparison. *Waste Management*, **33(1)**, pp. 117–122, 2013.
[16] Da Cruz, N.F., Simões, P. & Marques, R.C., Economic cost recovery in the recycling of packaging waste: The case of Portugal. *Journal of Cleaner Production,* **37**, pp. 8-18, 2012.
[17] Rada, E.C., Ragazzi, M. & Fedrizzi, P., WEB-GIS oriented system viability for municipal solid waste collection optimization in developed and transient economies. *Waste Management*, **33(4)**, pp. 785–792, 2013.
[18] Ionescu, G., Rada, E.C., Ragazzi, M., Marculescu, C., Badea, A. & Apostol T., Integrated municipal solid waste scenario model using advanced pretreatment and waste to energy processes. *Energy Conversion and Management*, **76**, pp. 1083-1092, 2013.
[19] Rada, E.C. & Ragazzi, M., Selective collection as a pretreatment for indirect solid recovered fuel generation. *Waste Management,* **34(2)**, pp. 291-297, 2014.
[20] Colon, J., Mestre-Montserrat, M., Puig-Ventosa, I. & Sanchez, A., Performance of compostable baby used diapers in the composting process with the organic fraction of municipal solid waste. *Waste Management,* **33(5)**, pp. 1097-1103, 2013.
[21] http://www.adep.provincia.tn.it
[22] Rada, E.C. & Andreottola, G., RDF/SRF: which perspectives for its future in the EU. *Waste Management*, **32(6)**, pp. 1059-1060, 2012.
[23] Ragazzi, M. & Rada, E.C., RDF/SRF evolution and MSW bio-drying. *WIT Transaction on Ecology and the Environment*, **163**, pp. 199-208, 2012.
[24] Velis, C.A. & Cooper, J., Are solid recovered fuels resource-efficient? *Waste Management and Research,* **31(2)**, pp. 113-114, 2013.
[25] Velis, C., Wagland, S., Longhurst, P., Pollard, S., Robson, B., Sinfield, K. & Wise, S., Solid recovered fuel: Influence of waste stream composition and processing on chlorine content and fuel quality. *Environmental Science and Technology,* **46(3)**, pp. 1923-1931, 2012.

Author index

www.ingramcontent.com/pod-product-compliance
Ingram Content Group UK Ltd.
Pitfield, Milton Keynes, MK11 3LW, UK
UKHW020802290926
15470UKWH00003B/919

9 781784 660604